T0335679

Earth Systems Data and Models

Volume 2

Series editors

Bernd Blasius, Carl von Ossietzky University Oldenburg, Oldenburg, Germany
William Lahoz, NILU—Norwegian Institute for Air Research, Kjeller, Norway
Dimitri P. Solomatine, UNESCO—IHE Institute for Water Education, Delft,
The Netherlands

Aims and Scope

The book series Earth Systems Data and Models publishes state-of-the-art research and technologies aimed at understanding processes and interactions in the earth system. A special emphasis is given to theory, methods, and tools used in earth, planetary and environmental sciences for: modeling, observation and analysis; data generation, assimilation and visualization; forecasting and simulation; and optimization. Topics in the series include but are not limited to: numerical, data-driven and agent-based modeling of the earth system; uncertainty analysis of models; geodynamic simulations, climate change, weather forecasting, hydroinformatics, and complex ecological models; model evaluation for decision-making processes and other earth science applications; and remote sensing and GIS technology.

The series publishes monographs, edited volumes and selected conference proceedings addressing an interdisciplinary audience, which not only includes geologists, hydrologists, meteorologists, chemists, biologists and ecologists but also physicists, engineers and applied mathematicians, as well as policy makers who use model outputs as the basis of decision-making processes.

More information about this series at http://www.springer.com/series/10525

Andrew Gettelman · Richard B. Rood

Demystifying Climate Models

A Users Guide to Earth System Models

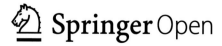

Andrew Gettelman
National Center for Atmospheric Research
Boulder
USA

Richard B. Rood
Climate and Space Sciences and Engineering
University of Michigan
Ann Arbor
USA

ISSN 2364-5830   ISSN 2364-5849   (electronic)
Earth Systems Data and Models
ISBN 978-3-662-48957-4   ISBN 978-3-662-48959-8   (eBook)
DOI 10.1007/978-3-662-48959-8

Library of Congress Control Number: 2015958748

Printed on acid-free paper

This Springer imprint is published by SpringerNature
The registered company is Springer-Verlag GmbH Berlin Heidelberg

Acknowledgments

Amy Marks provided very careful and thorough edit, as well as numerous helpful suggestions. Cheryl Craig, Teresa Foster, Andrew Dolan, and Galia Guentchev contributed their time to reading through drafts and providing a needed reality check. Prof. Reto Knutti helped this book take shape while Andrew Gettelman was on sabbatical at ETH in Zurich. David Lawrence shared critical insights and PowerPoint figures on terrestrial systems. Thanks also to Markus Jochum for straightening us out on explaining how the ocean works. Jan Sedlacek, ETH-Zürich, helped with figures in Chap. 11 (especially Fig. 11.6).

Mike Moran and David Edwards of the National Center for Atmospheric Research provided financial support. Lawrence Buja and the National Center for Atmospheric Research hosted Richard Rood's visitor status. The National Center for Atmospheric Research is funded by the U.S. National Science Foundation.

We thank the staff and students of the University of Michigan's Climate Center for reviews of the manuscript: Samantha Basile, William Baule, Matt Bishop, Laura Briley, Daniel Brown, Kimberly Channell, Omar Gates, and Elizabeth Gibbons. Richard Rood thanks the students in his classes on climate change problem-solving at the University of Michigan and acknowledges in particular the project work of: James Arnott, Christopher Curtis, Kevin Kacan, Kazuki Ito, Benjamin Lowden, Sabrina Shuman, Kelsey Stadnikia, Anthony Torres, Zifan Yang.

Richard Rood acknowledges the support of the University of Michigan and the Graham Sustainability Institute, and grants from the National Oceanographic and Atmospheric Administration (Great Lakes Sciences and Assessments Center (GLISA)—NOAA Climate Program Office NA10OAR4310213) and the Department of the Interior, National Park Service (Cooperative Agreement P14AC00898).

Francesca Gettelman exhibited nearly unlimited patience with some late nights.

Contents

About the Authors

Andrew Gettelman is a Scientist in the Climate and Global Dynamics and Atmospheric Chemistry and Modeling Laboratories at the National Center for Atmospheric Research (NCAR). He is actively involved in developing atmosphere and chemistry components for global climate models at NCAR. Dr. Gettelman specializes in understanding and simulating cloud processes and their impact on climate, especially ice clouds. He has numerous publications on cloud physics representations in global models, as well as research on climate forcing and feedbacks. He has participated in several international assessments of climate models, particularly for assessing atmospheric chemistry. Gettelman holds a doctorate in Atmospheric Science from the University of Washington, Seattle. He is a recent recipient of the American Geophysical Union Ascent Award, and is a Thompson-Reuters Highly Cited Researcher.

Richard B. Rood is a Professor in the Department of Climate and Space Sciences and Engineering (CLaSP) at the University of Michigan. He is also appointed in the School of Natural Resources and Environment. Prior to joining the University of Michigan, he worked in modeling and high performance computing at the National Aeronautics and Space Administration (NASA). His recent research is focused on the usability of climate knowledge and data in management planning and practice. He has started classes in climate-change problem solving, climate change uncertainty in decision making, climate-change informatics (with Paul Edwards). In addition to publications on numerical models, his recent publications include software engineering, informatics, political science, social science, forestry and public health. Rood's professional degree is in Meteorology from Florida State University. He recently served on the National Academy of Sciences Committee on A National Strategy for Advancing Climate Modeling. He writes expert blogs on climate change science and problem solving for the Weather Underground Richard Rood is a Fellow of American Meteorological Society and a winner of the World Meteorological Organization's Norbert Gerbier Award.

Introduction

Human-caused climate change is perhaps the defining environmental issue of the early twenty-first century. We observe the earth's climate in the present, but observations of future climate are not available yet. So in order to predict the future, we rely on simulation models to predict future climate.

This book is designed to be a guide to climate simulation and prediction for the non-specialist and an entry point for understanding uncertainties in climate models. The goal is not to be simply a popular guide to climate modeling and prediction, but to help those using climate models to understand the results. This book provides background on the earth's climate system and how it might change, a detailed qualitative analysis of how climate models are constructed, and a discussion of model results and the uncertainty inherent in those results. Throughout the text, terms in **bold** will be referenced in the glossary. References are provided as footnotes in each chapter.

Who uses climate models? Climate model users are practitioners in many fields who desire to incorporate information about climate and climate change into planning and management decisions. Users may be scientists and engineers in fields such as ecosystems or water resources. These scientists are familiar with models and the roles of models in natural science. In other cases, the practitioners are engineers, urban planners, epidemiologists, or architects. Though not necessarily familiar with models of natural science, experts in these fields use quantitative information for decision-making. These experts are potential users of climate models. We hope in the end that by understanding climate models and their uncertainties, the reader will understand how climate models are constructed to represent the earth's climate system. The book is intended to help the reader become a more competent interpreter or translator of climate model output.

Climate is best thought of as the distribution of weather states, or the probability of finding a particular *weather* state (usually described by temperature and precipitation) at any place and time. Climate science seeks to be able to describe this distribution. In contrast, the goal of predicting the weather is to figure out exactly which weather state will occur for a specific place and time (e.g., what the high temperature and total precipitation will be on Tuesday for a given city). Even in modern societies, we are still more dependent on the weather than we like to admit. Think of a winter storm snarling traffic and closing schools. Windstorms and hailstorms can cause significant damage. Or think of the impact of severe **tropical cyclones** (also called hurricanes or typhoons, depending on their location), personified and immortalized with names like Sandy, Andrew, or Katrina. Persistence (or absence) of weather events is also important. Too little rain (leading to drought and its resulting effects on agriculture and even contributing to wildfires) and too much rain (leading to flooding) are both damaging.

Although we are tempted to speak of a single "climate," there are many climates. Every place has its own. We build our societies to be comfortable during the expected weather events (the climate) in each place. Naturally, different climates mean different expected weather events, and our societies adapt. Buildings in Minneapolis are built to standards different from buildings in Miami or San Francisco. City planning is also different in different climates. Minneapolis has connected buildings so that people do not need to walk outside in winter, for example. Singapore has connected buildings so that people do not need to walk outside in heat and humidity. Not just the built environment, but the fabric of society may be different with local climates. In warmer climates, social life takes place outdoors, for example, or the flow of a day includes a rest period (*siesta* in Spanish) during the hottest period of the day.

We construct our lives for possible and sometimes rare weather events: putting on snow tires for winter even though snow may not be around for over half a winter. We build into our lives the ability to deal with variations in the weather. A closet contains coats, gloves, hats, rain jackets, umbrellas, sun hats, and sunglasses: We are ready for a range and for a distribution of possible weather states. Some events are rare: Snow occasionally has fallen in Los Angeles, for example. But it is usually the rare or extreme weather events that are damaging. These outliers of the distribution are typically damaging because they are unexpected and therefore we do not adequately prepare for them. Or rather, the expectation (probability) is so rare that it is not cost-effective for society to prepare for them. This applies to the individual as well: if you live in Miami your closet contains more warm weather gear, and less cold weather gear. It is unlikely you would be able to dress for temperatures well below freezing. The impacts of extreme weather are dependent on the climate of a place as well. For example, a few inches (centimeters) of snow is typical for Denver, Minneapolis, or Oslo, but it will shut down Rome or Atlanta. One inch (25 millimeters) of annual rainfall is typical for Cairo, but a disaster in most other places.

Where the most damaging weather events occur are at the extremes of the climate distribution. One problem is that we often do not know the distribution very well. Every time we hit a record (e.g., a high temperature, rainfall in 24 hours, days without rain), we expand the range of observed events a little, and we learn more about what might happen in a particular place. Because extreme events are rare, we do not really know the true chance of their occurrence. Think about your knowledge of the climate where you live or in a place you have visited several times. At first, you might not have a good grasp of what the seasons are like. After a few years, you think you know how the seasons evolve. But there are always events that will surprise you. The record events are those that surprise *everyone*. What is the probability of a hurricane flooding Houston as New Orleans was flooded by Katrina? It has not happened since the city of Houston has been there, so we may not know. The extremes of the distribution of possible weather states are not well known.

This creates even more of a problem when these extremes change. Changing the distribution of weather is what we mean by climate change. The cause of those changes might be natural or they might be human caused (anthropogenic).

So how do we predict the future of weather and the distribution of weather that represents climate in a location? To understand and predict the future, we need a way to represent the system. In other words, we need a model. This book is about how we attempt to use models to represent the complex climate system and predict the future. Our goal is to explain and provide a better understanding of the models we use to describe the past, present, and future of the earth system. These are commonly known as *climate models*. Scientists often refer to these models formally as earth system models, but we use the term *climate model*.

The purpose of this book is to demystify the models we use to simulate present and future climate. We explain how the models are constructed, why they are uncertain, and what level of confidence we should place in those models. Uncertainty is not a weakness. Understanding uncertainty is a strength, and a key part of using any model, including climate models. One key message is that the level of confidence depends on the questions we ask. What are we certain about in the future and why? What are we less certain of and why? For policy-makers, this is a critical issue. Understanding how climate models work and how we get there is an important step in making intelligent decisions using (or not using) these climate models. Climate models are being used not just to understand the earth system but also to provide input for policy decisions to address human-caused climate change. The direction of our environment and economy is dependent on policy options chosen based on results of these models.

The chapters in Part I serve as a basic primer on climate and climate change. We hope to give readers an appreciation for the complexity and even beauty of the complex earth system so that they can better understand how we simulate it. In Part II, we discuss the mechanics of models of the earth system: How they are built and what they are trying to represent. Models are built to simulate each region of the climate system (e.g., atmosphere, ocean, and land), critical processes within each region, as well as critical interactions between regions and processes. Finally, in Part III, we focus on uncertainties and probabilities in prediction, with a focus on understanding what is known, what is unknown, and the degree of certainty. We also discuss how climate models are evaluated. In the concluding chapter, we discuss what we know, what we may learn in the future, and why we should (or should not) use models.

Part I
Basic Principles and the Problem of Climate Forecasts

Chapter 1
Key Concepts in Climate Modeling

In order to describe climate modeling and the climate system, it is necessary to have a common conception of exactly what we are trying to simulate, and what a model actually is. What is climate? What is a model? How do we measure the uncertainty in a model? This chapter introduces some key terms and concepts. We start with some basic definitions of climate and weather. Everyone will come to this book with a preconceived definition of what climate and weather are, but separating these concepts is important for understanding how modeling of climate and weather are similar and why they are different. It also makes sense to discuss what a *model* is. Even if we do not realize it, we use models all the time. So we describe a few different conceptual types of models and put climate models in context. Finally we introduce the concept of uncertainty. As we discuss later in the book, models may have errors and still be useful, but this requires understanding the errors (the uncertainties) and understanding where they come from. Most of these concepts are common to many types of modeling, and we provide examples throughout the text.

1.1   What Is Climate?

Climate is perhaps easiest to explain as the distribution of possible **weather** states. On any given day and in any given place, the history of weather events can be compiled into a distribution with probabilities of what the weather might be (see Fig. 1.1). This figure is called a probability distribution function, representing a **probability distribution**.[1] The horizontal axis represents a value (e.g., temperature), and the vertical axis represents the probability (or frequency of occurrence) of that event's (i.e., a given temperature) occurring or having occurred. If based on observations, then the frequency can be the number of times a given temperature occurs. The higher the line, the more probable the event. The most frequent

---

[1]There is quite a bit of statistics in climate, by definition. For a technical background, a good reference is Devore, J. L. (2011). *Probability and Statistics for Engineering and the Sciences*, 8th ed. Duxbury, MA: Duxbury Press. Any specific aspect of statistics (e.g., standard deviation, probability distribution function) can be looked up on Wikipedia (www.wikipedia.com).

© The Author(s) 2016
A. Gettelman and R.B. Rood, *Demystifying Climate Models*,
Earth Systems Data and Models 2, DOI 10.1007/978-3-662-48959-8_1

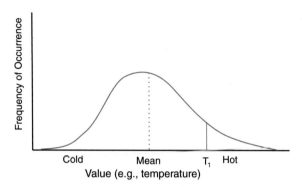

Fig. 1.1 A probability distribution function with the *value* on the horizontal axis, and the *frequency of occurrence* on the vertical axis

occurrence is the highest probability (the **mode**). The total area under the line is the *probability*. If the total area is given a value of 1, then the area under each part of the curve is the fractional chance that an event exceeding some threshold will occur. In Fig. 1.1, the area to the right of point T_1 is the probability that the temperature will be greater than T_1, which might be about 20 % of the curve. The **mean**, or expected value, is the weighted average of the points. It need not be the point with the highest frequency. The **mean** value is the point at which half the probability (50 %) is on one side of the mean, and half on the other. The **median** is the value at which half the points are on one side and half on the other.

Here is an important and obvious question: How can we predict the climate (for next season, next year, or 50 years from now) if we cannot predict the weather (in 5 or 10 days)? The answer is, we use probability: The climate is the distribution of probable weather. The weather is a particular location in that distribution, and it is conditional on the current state of the system. The chance of a hurricane hitting Miami next week depends mostly on whether one has formed or is forming, and if one has formed, whether it is heading in the direction of Miami. As another example, the chance of having a rainy day in Seattle in January is high. But, given a particular day in January, with a weather state that might be pushing storms well to the north or south, the probability of rain the next day might be very low. In 50 Januaries, though, the probability of rain would be high. So climate is the distribution of weather (sometimes unknown). Weather is a given state in that distribution (often uncertain).

In a probability distribution of climate, the probabilities and the curve change over time: In the middle latitudes, the chance of snow is higher in winter than in summer. The curves will look different from place to place: Some climates have narrow distributions (see Fig. 1.2a), which means the weather is often very close to the average. Think about Hawaii, where the average of the daily highs and lows do not change much over the course of the year or Alaska, where the daily highs and lows may be the same as in Hawaii in summer, but not in winter. For Alaska the annual distribution of temperature is a very broad distribution (more like Fig. 1.2b).

Fig. 1.2 Different probability
distribution functions:
a narrow, **b** wide and
c skewed distributions

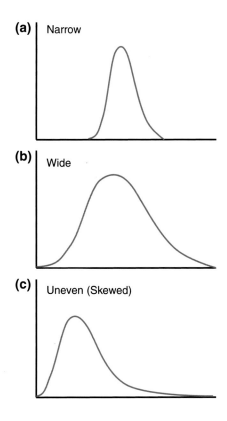

Of course, even in Hawaii, extreme events occur. For a distribution like precipitation, which is bounded at one end by zero (no precipitation) the distribution might
be "skewed" (Fig. 1.2c) with a low frequency of high events marking the
'extreme'. As events are more extreme (think about hurricanes like Katrina or
Sandy), there are fewer such events in the historical record. There may even be
possible extreme events that have not occurred. So our description of climate is
incomplete or uncertain. This is particularly true for rare (low-probability) events.
These events are also the events that cause the most damage.

One aspect of shifting distributions is that extremes can change a lot more than
the **mean** value (see Fig. 1.3). The mean is the value at which the area is equal on
each side of the distribution. The mean is the same as the median if the distribution
is symmetric. Simply moving the distribution to the right or left causes the area
(meaning, the probability) beyond some fixed threshold to increase (or decrease). If
the curve represents temperature, then shifting it to warmer temperatures (Fig. 1.3a)
decreases the chance of cold events and really increases the chance of warm events.
But note that some cold events still occur. Also, you can change the distribution
without changing the mean by making the distribution wider (or broader). The
mathematical term for the width of a distribution is **variance**, a statistical term for
variability. This situation is illustrated in Fig. 1.3b. The mean is unchanged in

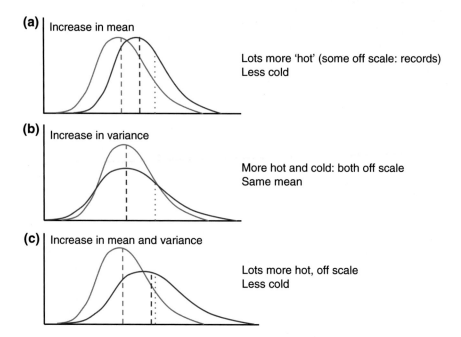

Fig. 1.3 Shifting probability distribution functions are illustrated in different ways going from the blue to red distribution. The thick lines are the distribution, the thin dashed lines are the mean of the distributions and the dotted lines are fixed points to illustrate probability. Shown is **a** increase in mean, **b** increase in variance (*width*), **c** increase in mean and variance

Fig. 1.3b, but the chance of exceeding a given threshold for warm or cold temperatures changes. In other words, the climate (particularly a climate extreme) changes dramatically, even if the mean stays the same. The change need not be symmetric: Hot may change more than cold (or vice versa). Figure 1.3c is not a symmetric distribution. The key is to see climate as the distribution, not as a fixed number (often the mean).

This brings us to the fundamental difference between weather and climate forecasting. In **weather forecasting**,[2] we need to know the current state of the system and have a model for projecting it forward. Often the model can be simple. One "model" we all use is called *persistence*: What is the weather now? It may be like that tomorrow. In many places (e.g., Hawaii), such a model is not bad, but sometimes it is horribly wrong (e.g., when a hurricane hits Hawaii). So we try to use more sophisticated models, now typically numerical ones. These models go by the name **numerical weather prediction (NWP) models**, and they are used to "forecast" the evolution of the earth system from its current state.

---

[2]For an overview of the history of weather forecasting, see Edwards, P. N. (2013). *A Vast Machine: Computer Models, Climate Data, and the Politics of Global Warming*. Cambridge, MA: MIT Press.

Climate forecasting uses essentially the same type of model. But the goal of climate forecasting is to characterize the distribution. This may mean running the model for a long time to describe the distribution of all possible weather states correctly. If you start up a weather model from two different states (two different days), you hope to get a different answer each time. But for climate, you want to get the same distribution (the same climate), regardless of when the model started. We return to these examples again later.

1.2 What Is a Model?

A **model**, in essence, is a representation of a system. A model can be physical (building blocks) or abstract (an image on paper like a plan, or in your head). Abstract models can also be mathematical (monetary or physical totals in a spreadsheet). Ordinary physics that describes how cars go (or, more importantly, how they stop suddenly) is a model for how the physical world behaves. Numbers themselves are abstract models. A financial statement is a model of the money and resource flows of a household, corporation or country. Models are all around us, and we use them to abstract, make tractable and understand our human and natural environment.

As a concrete example, think about different "models" of a building. There can be many types of models of a building. A physical model of the building would usually be at a smaller scale that you can hold in your hand. There are several different abstract models of a building, and they are used for different purposes. Architects and engineers produce building plans: two-dimensional representations of the building, used to construct and document the building. Some of these are highly detailed drawings of specific parts of the building, such as the exterior, or the electrical and plumbing systems. The engineer may have built not just a physical model of the building, but perhaps even a more detailed structural model designed to understand how the building will react to wind or ground motion (earthquakes). Increasingly, these models are "virtual": The structure is simulated on a computer.

We are familiar with all these sorts of model, but there are other more abstract models that deal with flows and budgets of materials or money. The owner and builder also probably have a spreadsheet model of the costs of construction of the building. This financial model is not certain, because it is really an estimate (or forecast, see below) of all the different costs of construction. And, finally, the owner likely has another model of the financial operation of the building: the money borrowed to finance the building, any income from a commercial building, and the costs of maintenance and operations of the building. The operating plan is really a projection into the future: It depends on a lot of uncertainties, like the cost of electricity or the value of the income from a building. The projection depends on these inputs, which the spreadsheet does not try to predict. The prediction is conditional on the inputs.

Models All Around Us

Models are everywhere in our world. Many models are familiar and physical, such as a small-scale model of a building or a bridge, or a mockup of a satellite or an airplane. Some models we use every day are made up of numbers. Many people use a model with numbers to manage income and expenses, savings and debts; that is, a budget. When the model of a bridge is placed in a computer-assisted design program, or a financial budget is put into a spreadsheet on a personal computer, then one has a "numerical" model. These models have a set of mathematical equations that behave with a specific set of rules, principles or laws.

Climate models are numerical models that calculate budgets of mass, momentum (velocity) and energy based on the physical laws of conservation. For example, energy is conserved (neither created or destroyed) and can, therefore, be counted. The physical laws on which climate models are based are discussed in more detail in Chap. 4. Weather and climate are dynamical systems; that is, they evolve over time. We rely on models of dynamical systems for many aspects of modern life. Here we illustrate a few examples of models that affect our everyday lives.

Climate models are closely related to weather forecast models; both simulate how fluids (air or water) move and interact, and how they exchange heat. An obvious example of a model that affects daily life is the weather forecast model, which is used in planning by individuals, governments, corporations and finance. The exact same physical principles of fluid flow are used to simulate a process in a chemical plant that takes different substances as liquids or gasses, reacts them together under controlled temperature and pressure, and produces new substances. Water and sewage treatment plants share similar principles and models. Internal combustion engines used for cars, trucks, ships and power plants are developed using models to understand how fuel enters the engine and produces heat, and that heat produces motion. Airplanes are also developed using modeling of the airflow around an aircraft. In this case, computational modeling has largely replaced design using wind tunnels. All of these models involve fluid flow and share physical principles with climate models. The details of the problem, for example, flow in pipes as contrasted to flow in the free atmosphere, define the specific requirements for the model construction.

So do you trust a model? Intuitively, we trust models all the time. You are using the results of a model every time you start your car, flush your toilet, turn on a light switch or get in an airplane. You count on models when you drive over a bridge. When NASA sends a satellite or rover to another planet, the path and behavior of the space vehicle relies on a model of simple physics describing complex systems.

Models do not just describe physical objects. Models of infectious diseases played an important role in management of the 2014 outbreak of Ebola. One function of models is to provide plausible representations of events to come, and then to place people into those plausible futures. It is a way to anticipate and manage complexity. Though most times these models do not give an exact story of the future, the planning and decision making that comes from these modeling exercises improves our ability to anticipate the unexpected and to manage risk. Think of modeling as a virtual, computational world in which to exercise the practice of trial and error, and therefore, a method for reducing the "error" in trial and error. Reducing these errors saves lives and property. Models reduce the chance of errors: Airplanes do not regularly fall out of the sky, bridges do not normally collapse and chemical plants do not typically leak.

Ultimately, trust of models is anchored in evaluation of models compared to observations and experiences. Weather models are evaluated every day with billions of observations as well as billions of individuals' experiences. Trust is often highly personal. By many objective measures, weather models have remarkable accuracy, for example, letting a city know more than five days in advance that a major tropical cyclone is likely to make landfall near that city. Of course, if the tropical cyclone makes landfall just 60 miles (100 km) away from the city, many people might conclude the models cannot be trusted.

Objectively, however, a model that simulated the tropical cyclone and represented its evolution with an error of 60 miles (100 km) on a globe that spans many thousands of miles can, also, be construed as being quite accurate. This represents the fact that models provide plausible futures that inform decision making.

Like weather models, climate models are evaluated with billions of observations and investigations of past events. The results of models have been scrutinized by thousands of scientists and practitioners. With virtual certainty, we know the Earth will warm, sea level will rise, ice will melt and weather will change. They provide plausible futures, not prescribed futures. There are uncertainties, and there will always be uncertainties. However, our growing experiences and vigilant efforts to evaluate and improve will help us to understand, manage and, sometimes, reduce uncertainty. As the models improve, trust and usability increase. There remains some uncertainty in most physical models, but that can be accounted for, and we discuss uncertainty, and its value in modeling, at length.

Our world is completely dependent on physical models, and their success is seen around us in the fact that much of the world "works" nearly all of the time. Models are certain enough to use in dangerous contexts that are both mundane and ubiquitous. We answer the question of whether we should trust models and make changes in our lives based on their results every time we get in an elevator or an airplane. There is no issue of should we trust models; we have been doing it for centuries since the first bridge was constructed, the first train left a station or the first time a building was built more than one story high.

1.3   Uncertainty

Forecasting involves projecting what we know, using a model, onto what we do not know. The result is a prediction or forecast. Forecasts may be wrong, of course, and the chance of them being wrong is known as **uncertainty**. Uncertainty can come from several different sources, but this is particularly the case when we think about climate and weather. One way to better characterize uncertainty is to divide it into categories based on model, scenario and initial conditions.[3]

1.3.1   Model Uncertainty

Obviously, a model can be wrong or have structural errors (**model uncertainty**). For example, if one were modeling how many tires a delivery company would need for their trucks in a year and assumed that the tires last 10,000 miles, when they actually only last 7,000 miles, the *forecast* of tire use is probably wrong. If you assumed each truck would drive 20,000 miles per year and tires last 10,000 miles, the trucks would need two sets of tires. However, what if the trucks drove only 14,000 miles per year and each set of tires lasted only 7,000 miles, (still needing just two sets of tires)? Then the forecast might be right, but the tire forecast would be right for the wrong reason: in this case, a *cancellation of errors*.

1.3.2   Scenario Uncertainty

The preceding example also illustrates another potential uncertainty faced in climate modeling: **scenario uncertainty**. The scenario[4] is the uncertainty in the future model inputs. In the tire forecast, the scenario assumed 20,000 miles per truck each year. But the scenario was wrong. If the tire forecast model was correct (or "perfect") and tires lasted 10,000 miles, but the mileage was incorrect (14,000 vs. 20,000 assumed), then the forecast will still be incorrect, even if the model is perfect. If the actual mileage continued to deviate from the assumption (14,000 miles), then the forecast over time will continue to be incorrect. If one is concerned with the total purchase of tires and total cost, then the situation becomes even more uncertain. Other factors (e.g., growth of the company, change in type of tires) may make forecasting the scenario, or inputs to the model, even more uncertain, even if the model as it stands is perfect. As the timescale of the model looks farther into the

---

[3]This definition of uncertainty has been developed by Hawkins, E., & Sutton, R. (2009). "The Potential to Narrow Uncertainty in Regional Climate Prediction." *Bulletin of the American Meteorological Society, 90*(8): 1095–1107.

[4]For a discussion of climate scenarios, see Chap. 10.

future, more and more different "variables" become uncertain (e.g., new types of tire, new trucks, the cost of tires). These variables that cannot be predicted, but have to be assumed, are often called **parameters**. Scenario uncertainty logically dominates uncertainty farther into the future (see Chap. 10 for more detail).

1.3.3 Initial Condition Uncertainty

Finally, there is uncertainty in the initial state of the system used in the model, or **initial condition uncertainty**. In our tire forecast, to be specific about how many tires we will need in the current or next year, we also need to know what the current state of tires is on all the trucks. Changes in the current state of tires will have big effects on the near-term forecast: If all trucks have 6,000 or 8,000 miles on their tires, there will be more purchases of tires in a given year than if they are all brand new. Initial condition uncertainty is a similar problem in weather forecasting. As we will discuss, some aspects of the climate system, particularly related to the oceans, for example, have very long timescales and "memory," so that knowing the state of the oceans affects climate over several decades. However, over long timescales (longer than the timescale of a process), these uncertainties fade. If you want to know how many tires will be needed over 5 or 10 years, the uncertainty about the current state of tires (which affects only the first set of replacements) on the total number of tires needed is small.

1.3.4 Total Uncertainty

In climate prediction, we must address all three types of uncertainties—model uncertainty, scenario uncertainty and initial condition uncertainty—to estimate the total uncertainty in a forecast. They operate on different time periods: Initial condition uncertainty matters most for the short term (i.e., weather scales, or even seasonal to annual, in some cases), and scenario uncertainty matters most in the longer term (decades to centuries). Model uncertainty operates at all timescales and can be "masked" or hidden by other uncertainties.

The complicated nature of these uncertainties makes prediction both harder and easier. It certainly makes it easier to understand and characterize the uncertainty in a forecast. One of our goals is to set down ideas and a framework for understanding how climate predictions can be used. Judging the quality of a prediction is based on understanding what the uncertainty is and where it comes from. Some comes from the model and some comes from how the experiment is set up (the initial conditions and the scenario).

1.4   Summary

Climate can best be thought of as a distribution of all possible weather states. What matters to us is the shape of the distribution. Weather is where we are on the distribution at any point in time. The extreme values in the distribution, that usually have low probability, are hard to predict, but that is where most of the impacts lie. Weather and climate models are similar, except weather models are designed to predict the exact location on a distribution, while climate models describe the distribution itself.

We use models all the time to predict the future. Some models are physical objects, some are numerical models. Climate models are one type of a numerical model: As we shall see, they can often be thought of as giant spreadsheets that keep track of the physical properties of the earth system, the same way a budget keeps track of money.

Uncertainty in climate models has several components. They are related to the model itself, to the initial conditions for the model (the starting point) and to the inputs that affect the model over time in a "scenario." All three must be addressed for a model to be useful. Uncertainty is not to be feared. Uncertainty is not a failure of models. Uncertainty can be understood and used to assess confidence in predictions.

Key Points

- Climate is the distribution of possible weather states at any place and time.
- Extremes of climate are where the impacts are.
- We use models all the time to predict the future, some models are even numerical.
- Weather and climate models are similar but have different goals.
- Uncertainty has several different parts (model, scenario, initial conditions).

Chapter 2
Components of the Climate System

We experience climate generally at the surface of the earth. This is the intersection of a number of different and distinct parts of the climate system. Understanding the different components of the climate system is critical for being able to simulate the system. As we will discuss, the climate system is typically simulated as a set of building blocks, from each individual process (i.e., the condensation of water to form clouds) collected into a model of one part or component of the system (i.e., the atmosphere), and then coupled to other components of the system (i.e., ocean, land, ice). Understanding and then representing in a model the different interactions between processes and then between components is critical for being able to build a representation of the system: a climate model.

In this chapter we describe the basic parts of the earth system that comprise the climate system, some of the key scientific principles and critical processes necessary to simulate each of these components. This forms the background to a discussion of how climate might change (see Chap. 3) and a more detailed discussion of each component and how it is simulated (Sect. 2.2, Chaps. 5–8). We discuss the key components of the earth system, as well as some of the critical interactions (discussed in more detail in Chap. 8).

2.1 Components of the Earth System

Figure 2.1 represents a schematic of many of the important components of the earth system that govern and regulate climate. Broadly, there are three different regions of the planet: the atmosphere, the oceans and the land (or terrestrial) surface. In addition to these general regions, we also speak of a cryosphere, the snow and ice covered regions of the planet. This fourth "sphere" spans the ocean (as sea ice) and the terrestrial surface (as glaciers, snow and ice sheets). We address the cryosphere in discussions of the ocean and terrestrial surface. While modeling the surface of the earth is commonly thought of as just modeling the land surface, it also includes the cryosphere (ice and snow) that sits on land. The term *terrestrial* is used to encompass all these spheres, though the common term *land* is also used.

© The Author(s) 2016
A. Gettelman and R.B. Rood, *Demystifying Climate Models*,
Earth Systems Data and Models 2, DOI 10.1007/978-3-662-48959-8_2

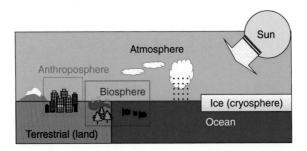

Fig. 2.1 The Earth system. The climate system contains different spheres (components): atmosphere, ocean, terrestrial, cryosphere, biosphere and anthroposphere

These are the traditional physical components of the earth's climate system. We also introduce two more "spheres." An important fifth component of the system is the biosphere: the living organisms on the planet, again, which span the terrestrial surface (plants, organisms in the soil, and animals) as well as the ocean (fish and plants in the ocean). We discuss the biosphere as part of both the ocean and terrestrial surface. Finally, although humans are technically part of the biosphere, our large "footprint" and impact on the global environment and the climate system is large enough that we can define a separate sphere for human activity and impacts called the *anthroposphere* (see Chap. 3).

2.1.1 The Atmosphere

The **atmosphere** is usually the first part of the climate system we naturally think of. It is literally the air we breathe: mostly inert nitrogen (78 %) with oxygen (16 %) and then other trace gases (argon, water vapor, carbon dioxide). The oxygen is a by-product of the respiration ("breathing") of plants and other organisms: It is evidence of life on earth. The oxygen in the atmosphere did not exist before the emergence of living organisms.[1] Oxygen is emitted by plants as an outcome of photosynthesis that removes carbon from carbon dioxide. Oxygen reacts with materials (rock and ore) at the earth's surface (oxidation) and disappears from the atmosphere. One of the most common reactants is iron (iron oxide = rust), which is responsible for the red color of many rocks. Unless organisms continue to produce oxygen, it will disappear from the atmosphere. It would take a long time however: hundreds of thousands to millions of years. But it is the trace species—water vapor, carbon dioxide and methane—known as the **greenhouse gases**, that are most important in understanding the climate system and how climate might change.

---

[1]Kasting, J. F., & Siefert, J. L. (2002). "Life and the Evolution of Earth's Atmosphere." *Science*, *296*(5570): 1066–1068.

The term *greenhouse gas* refers to gases that are transparent to visible light emitted from the sun, but opaque (absorptive) to the "infrared" radiation emitted by the earth. This follows from fundamental physical principles described by the Boltzmann law, after Ludwig Boltzmann, a 19th-century Austrian physicist. All mass radiates energy, depending on temperature. The hotter a body (could be your body, the earth, or the sun), the more energy it radiates: and the peak energy occurs at different wavelengths. The peak radiation of the sun (surface temperature = 6,000 Kelvin[2] [K] or 10,000 °F: very hot either way) radiates a lot of the light we call visible. It is no accident our eyes evolved to be able to 'see' in the visible where the maximum solar emission is. Objects with cooler temperatures such as the temperature of the earth at about 300 Kelvin (80 °F) radiate in the infrared (longer wavelengths with much lower energy). These objects include our bodies, the earth itself, the ocean surface or clouds.

Greenhouse gases, like the glass in a greenhouse, allow in visible light from the sun, but absorb (and re-emit back down) the infrared energy from the earth.[3] The higher the concentration of these gases in the atmosphere, the greater the warming effect. The most important greenhouse gas is water vapor. Water vapor also has some other important effects in the climate system. The second most important gas is carbon dioxide.

The basic picture of the flows of energy in the atmosphere is illustrated in Fig. 2.2. The arrows represent approximate sizes of the **energy flows**[4] in the energy budget. Solar energy comes in; the atmosphere, clouds and the earth absorb some and reflect some. The technical term for the ratio of reflected over total energy is **albedo**. White/light surfaces reflect a lot and have a high albedo: snow, ice, bright sand. Black/dark surfaces absorb a lot and have a low albedo: dark green trees, the ocean, asphalt. This is why a black car is hotter than a white car or asphalt is hotter than concrete in the sun. The energy absorbed warms the object (clouds, atmosphere, surface of the earth/ice/ocean) and it 're-emits', but now at longer wavelengths and lower energy since the temperature of the surface is much lower than the sun. For common temperatures at the earth's surface (0–100 °F, –20 to 40 °C), this emission occurs in the infrared. Infrared radiation is absorbed in the atmosphere by greenhouse gases and clouds, and some returns to the earth. The earth warms the atmosphere above it in two ways. The direct conduction of heat from a warm

---

[2]The Kelvin temperature scale has the same increment as the Celsius scale but starts at absolute zero, while Celsius starts at the freezing point of water and can be negative. The Fahrenheit scale has a smaller increment and the freezing point of water is 32 °F. So 1 °K = 1 °C = 9/5 °F. Scientists use °K, the United States uses °F and most other countries use °C. We will typically provide °F and °C.

[3]Strictly speaking, the analogy is not correct: The glass in a greenhouse also prevents air from escaping, and simply being transparent in the visible and restricting air motion is sufficient to keep a greenhouse warm. The actual "greenhouse" effect for glass is a small part of it.

[4]For a more detailed treatment of the energy budget, see Trenberth, K. E., Fasullo, J. T., & Kiehl, J. (2009) "Earth's Global Energy Budget." *Bulletin of the American Meteorological Society, 90*(3): 311–323.

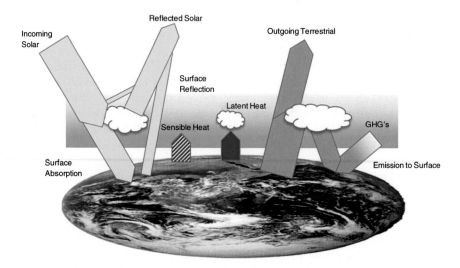

Fig. 2.2 Energy budget. Solar energy, or shortwave radiation (*yellow*) comes in from the sun. Energy is then either reflected by the surface or clouds or absorbed by the atmosphere or surface (mostly). The surface exchange includes sensible heat (*red striped*) and latent heat (*blue*) associated with water evaporation and condensation. Terrestrial (*infrared*, longwave) radiation (*purple*) emitted from the earth's surface is absorbed by the atmosphere and clouds. Some escapes to space (outgoing terrestrial) and some is re-emitted back to the surface by clouds and greenhouse gases (GHGs)

surface occurs by sensible heat. The energy hitting the surface, particularly the ocean surface, also can evaporate water vapor. Since this water vapor contains the energy used to evaporate it, it also is a way of transmitting 'heat' (energy), without changing the temperature. The energy in water vapor is called "latent heat".

This is the cycle of radiative energy in the earth system, flowing through the atmosphere. The ultimate source of the energy is the sun. One of the biggest complications is water, in all its forms: from the oceans, to water vapor in the atmosphere, to clouds. Put a cloud over a dark surface (ocean or forest), and you change the energy in the system. The energy changes by reflecting solar energy from the top of the cloud back to space (cooling). Then energy also changes by absorbing infrared energy radiated from below in the cloud and sending some of it back down to the surface (warming).

The movement of water through the atmosphere, the surface and the ocean is called the **hydrologic cycle** (Fig. 2.3), which is also important for moving energy in the atmosphere. Solar energy hitting the ocean causes **evaporation** of vapor into the atmosphere, and plants also move water from liquid phase in the soil through their leaves, back into the atmosphere in a process called **transpiration**. This water carries the energy necessary for evaporation in the form of latent heat. The energy is used to make water molecules in liquid move fast enough to separate from each other into a gas. They conserve this heat of evaporation as the water vapor moves through the atmosphere. This heat of evaporation is released when water vapor undergoes **condensation** into cloud drops (or ice crystals). This can happen a long way from the

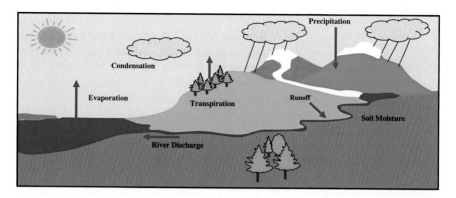

Fig. 2.3 The Hydrologic cycle. Water evaporates from the ocean and is carried in the atmosphere as vapor. It condenses into clouds, and precipitates to the surface, where it can become soil moisture, glaciers, or runoff as surface water into lakes and back to the ocean. Plants move soil water back to the atmosphere through transpiration

liquid source (oceans), and clouds eventually form enough drops that they get heavy and fall (precipitate). The water thus completes a cycle, landing back on the surface.

Water then enters other components of the earth system, such as glaciers, snow, soil or rivers. Soil moisture can go through plants to get back into the atmosphere by transpiration. Or it can saturate the soil, become runoff and eventually get back into the ocean, where the cycle repeats. The key is that water changes the energy budget because it forms clouds, because it is a greenhouse gas and because it moves heat around. Water evaporated in the tropics moves as water vapor with the winds to higher latitudes (see Chap. 5), where it condenses as clouds and deposits heat there. Water is magical stuff in the earth system. The cycle of water is linked to the energy budget (through latent heat), and it links the atmosphere to other pieces of the system.

2.1.2 The Ocean and Sea Ice

We have discussed the ocean briefly as a source of water. The world's oceans (or combining them into a single "ocean") are a critical part of the climate system.[5] The ocean is a tremendous reservoir of heat and holds a lot more mass than the atmosphere. Water is denser than air, and the mass of the top 30 feet (10 m) of ocean is equal to the mass of the atmosphere above it (90 % of the atmosphere is in the first 55,000 feet or about 17 km). The ocean is critical as a very large store of heat that can move around and come back into contact with the surface after long periods of time.

---

[5]For a good basic overview of the ocean and climate, see Vallis, G. K. (2012). *Climate and the Oceans*. Princeton, NJ: Princeton University Press.

The ocean has currents that are both shallow (near the surface, the first few hundred feet of depth) and deep (the rest of the ocean). Unlike the atmosphere, the ocean has "edges": Land sticks up in places, dividing the seven seas into, well, seven seas (actually five ocean basins). The ocean surface currents are forced by the rotation of the earth and winds. These currents are constrained by the topography in ocean basins and generate **circulation patterns** of surface currents.

In addition to the near surface ocean, there is a complex vertical structure to the ocean. The ocean is divided into a near surface region that is generally warmer and less salty (fresher) than the deeper ocean below it. Colder and saltier water is denser, so it tends to sink. The process often is self-reinforcing and is seen in lakes as well: The surface warms, gets less dense and mixes less with the deeper water, thus warming more over time (this is called stratification). Density is a critical part of the structure of the ocean because of changes in density due to heat and salinity, which do not change very quickly in the ocean. Density helps control the deep ocean circulation. The deep ocean has one large circulation globally, driven by the sinking of surface water that gets saltier and colder (denser) until it sinks from the surface to the deep ocean. This **thermohaline circulation** (*thermo* = heat, *haline* = salt) has sinking motion in small areas of the North Atlantic and the Southern Ocean near Antarctica, where water is cooling as it moves poleward, and where salt is expelled from formation of sea ice, increasing salinity. Both processes make the surface water denser than the water beneath it, causing it to sink.

The ocean also contains parts of two other spheres. The ocean has a biosphere, consisting of plants and animals living in the ocean and nutrients that flow through the ocean. Most of the biosphere in the ocean is composed of small organisms, mostly small floating plants (phytoplankton, algae) and animals that form the basis of the marine food chain. The ocean is an important part of the flows of carbon through the earth system as well. The ocean also holds (or supports) part of the cryosphere as sea ice.[6] Sea ice forms in winter in high latitudes in the Arctic Ocean and in the Southern Ocean. The sea ice moves around and can grow and melt. The ice may last several years and is a year-round feature in most of the Arctic today.

2.1.3 *Terrestrial Systems*

Most of the land surface is not covered by ice, however. Most of it is covered by plants: the **terrestrial biosphere**. There are also lakes and rivers that channel precipitation and return it to the ocean. The atmosphere and ocean cause changes over time in the composition of the land by erosion and processing of rock material. Water, wind, and waves break up rock. The rock reacts with the atmosphere (oxidation) and with water: Sometimes minerals in rocks dissolve in water. The land

---

[6]For a primer on sea ice, see Marshall, S. J. (2011). *The Cryosphere*. Princeton, NJ: Princeton University Press.

evolves by moving slowly as it floats over the crust. For purposes of climate, the land is often considered static, but we know it changes over long timescales. Volcanoes and earthquakes are evidence of this motion.

The terrestrial surface is intimately connected with the plants and the biosphere that lives on it: Dead plant (and a bit of dead animal) material makes up a significant part of the soil. Plants regulate the water in the soil through uptake into plant structures and release water into the atmosphere. Plants also can be darker or lighter than the surface soil, thus absorbing (and reflecting) a different amount of energy than in their absence. Plants react to their environment (rainfall), but plants and the biosphere also create their environment. Plants cycle water vapor back into the atmosphere from the soil by moving water from the soil into the plant, where it is lost in the process of photosynthesis back to the atmosphere. Plants also change the reflectance of the surface, altering the energy absorbed in the system, and the temperature of the surface. The biosphere cycles nutrients used by plants from the soil into plant material and back again, sometimes also into the ocean. Central to the biosphere are nutrients, and the elements that make up the structure of organic molecules: oxygen (O) and hydrogen (H) and carbon (C).

But some of the terrestrial surface, particularly at high latitudes, is frozen. This, of course, brings us to a discussion of ice and snow at high latitudes and altitudes: the **cryosphere**. Snow may be present seasonally on land at high latitudes or high altitudes. If snow lasts over the summer, and continues to pile up, it compacts into ice and forms glaciers. If glaciers get big enough, they merge into ice sheets, ranging in size from small ice sheets filling a few valleys in New Zealand or Alaska, to ice sheets covering entire land masses (Greenland) or continents (Antarctica). The Greenland and Antarctic ice sheets are up to 10,000 feet thick; 99 % of the fresh water on earth is in ice sheets. Greenland holds enough frozen water to raise the sea level 20 feet (6 m). Antarctica holds enough frozen water to raise the sea level 200 feet (60 m).[7]

Finally, while physically humans are part of the biosphere, our impact on the planet is large enough that they are often treated as a separate sphere, the **anthroposphere**. Human emissions as a by-product of our societies now play a significant role in the climate system. Humans alter the structure of ecosystems on continent wide scales (deforestation, and subsequent afforestation). By-products of our societies add to the greenhouse gas loading of the atmosphere, and change atmospheric chemical processes near the surface (creating extra ozone, which is bad for humans, plants and animals). In the upper atmosphere some of these compounds destroy ozone in the stratosphere necessary to block ultraviolet radiation from reaching the surface. These human actions are often modeled using economic models, or models of entire societies and their emissions. These models are sometimes now treated as part of the climate system because the outputs from societies can alter the climate system.

---

[7]Values are from the National Snow and Ice Data Center (NSIDC). http://nsidc.org/cryosphere/quickfacts/icesheets.html.

2.2   Timescales and Interactions

There is a huge variety of timescales to these cycles: from seconds to millions of years. Evaporation and precipitation formation take seconds to minutes to days to cycle water through the atmosphere, but water that falls onto ice sheets may remain there for hundreds of thousands of years. The energy budget changes throughout the course of the day as the sun moves across the sky and throughout the course of a year as the sun varies its position at any place. Many of the cycles are regulated by the motion of the earth relative to the sun. Plant growth and decay cycles follow the annual cycle. On even longer timescales of thousands to tens of thousands of years, the earth wobbles a bit on its tilted axis (and the tilt changes) and in its orbit. The wobbles alter the seasonal intensity of sunlight at any given location and also alter the distance between the earth and the sun. The result is a slow change in the amount of sunlight hitting the top of the atmosphere. These orbital cycles make it warmer or colder at higher latitudes, enhancing or retarding the presence of snow and ice on the ground, ultimately leading to ice ages.[8] The motion of the plates, the slow tectonic shifts that alter the terrestrial surface and expose new land over millions of years, will change the ocean basins, as well as changing the supply of raw carbon from rocks into the system, or the amount of volcanism that puts new rock onto the surface (and gases from the earth into the atmosphere).

The interplay of cycles such as flows of water and carbon alters many aspects of the climate system. Changing the amount of carbon dioxide in the atmosphere changes the greenhouse effect and the temperature. The temperature change alters water vapor concentrations, potentially further altering the temperature. The coupled nature of the earth's climate system gives rise to a constantly evolving series of interacting processes between the different components. The interplay of forces and pieces is a slowly evolving system with very different interactions and timescales. The climate system is a dynamical system, meaning it can be altered in many ways, and the climate will evolve or change over time. How it will change (or not change) is of critical importance to ourselves and society, and why we try to build models to understand it.

Some of the most critical interactions and flows in the climate system concern carbon, as carbon dioxide (CO_2), and water (H_2O). Both CO_2 and H_2O are critical for life and the biosphere. They are the building blocks of organic molecules, but CO_2 and H_2O are also part of the atmosphere, and make up the largest greenhouse gases, along with methane (CH_4). This is another example of the connections and interactions in the climate system. These connections are one of the reasons why climate change is hard to understand: Carbon dioxide is bad? Not always: Plants need it for their "breathing" (respiration). Carbon dioxide is used in photosynthesis, and for that more CO_2 makes photosynthesis more efficient. Water and carbon flow through the climate system and link the different components.

---

[8]Hays, J. D., Imbrie, J., & Shackleton, N. J. (1976). "Variations in the Earth's Orbit: Pacemaker of the Ice Ages." *Science*, *194*(4270): 1121–1132.

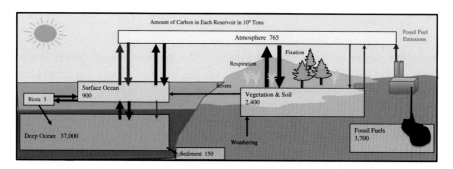

Fig. 2.4 The Carbon Cycle. The largest climate system reservoirs include the deep ocean, soil and vegetation, the surface ocean and the atmosphere. The approximate size of exchanges (fluxes) between boxes is given by the width of the arrows; red arrows indicate perturbations by humans

The cycle of carbon is illustrated in Fig. 2.4. Carbon dioxide in the atmosphere is a greenhouse gas that affects the energy budget, but it may also get taken up by plants (**fixation**) and form part of their structure (leaves, stems, roots). When the plant material dies, carbon enters the soil. Carbon may be broken down by microbes and get back into the atmosphere (**respiration**) as a carbon-containing gas (e.g., CH_4, also a greenhouse gas). It may be dissolved into water and carried into the ocean by rivers, where it also may get taken up into plant or animal tissue (oceanic biota), or remain in the water column. Oceanic biomass "rains" down to the deep ocean when parts of the oceanic biosphere die, and this carbon forms sediments on the bottom of the sea. Some of the carbon dissolves in the water column, where it comes into balance (**equilibrium**) with the atmosphere. The sediments under the ocean eventually get compacted by sediments above, and shifted around over millions of years by tectonic plate motions, where they come back onto land, and rock is exposed to the atmosphere, where it is broken up by wind and water (**weathering**), and the carbon can be dissolved again in water, and react to release the carbon back into the air. This completes a cycle that might take millions of years. The cycle moves carbon in the earth's climate system and through the biosphere. The carbon cycle contains carbon that makes up our living tissues: literally our blood.

As noted in Fig. 2.3, water flows throughout the climate system in the hydrologic cycle. Water evaporates from the ocean into the atmosphere, forming clouds and affecting atmospheric and surface energy transport. It may fall on the land as rain (or snow). Snow affects the energy budget by reflecting more sunlight to space. When snow melts, the water can go into rivers or the soil, where it can be taken up by plants, and then released to the atmosphere as a side effect of pores in leaves opening up to allow carbon from the atmosphere in for photosynthesis. The water may then form a cloud and precipitate, perhaps back into the ocean. The precipitated water is fresh water, lowering the salt content and density of the surface ocean, and affecting the ocean circulation.

2.3 Summary

The earth's climate system is driven by the sun. The atmosphere mediates the flow of energy between the sun and the earth, through the action of clouds and greenhouse gases. There are several different components of the earth's climate system, usually divided into "spheres": atmosphere, terrestrial surface, ocean, cryosphere, biosphere and even anthroposphere (the sphere of human effects). In addition to energy, several critical substances flow through the earth system. Two of the most important and unique are water and carbon. They are important for climate, and they are important for life: Almost all living things contain and use carbon in different forms for our bodies and as part of the cycle of energy (either photosynthesis in plants or respiration in animals).

Both carbon and water have important cycles in the earth's climate system. In the atmosphere they are greenhouse gases. Water is important for life and also as a mechanism for moving heat from where it evaporates (taking up the heat to evaporate water) and releasing it on condensation. Carbon is stored in soils and rocks, and dissolved as a gas in the ocean. We will learn more about the carbon cycle in Chap. 7. One of the reasons the climate system is complex, and comprehensive models are necessary, is because these greenhouse gases (carbon dioxide and water vapor) are critical for understanding the energy budget, but they also flow through the whole climate system.

Thus many interactions between parts of the climate system are critical for describing how climate works and how it evolves. The cycles work on many different timescales. The motion of the earth is a good example: It causes day and night, it causes the seasons, and variations in the earth's orbit over thousands of years alter the conditions at the earth's surface. The slow drift of continents and variations in weathering can alter the climate on even longer timescales. These pieces, and timescales, are critical for understanding how climate may change, the subject of Chap. 3.

Key Points

- The energy in the climate system comes from the sun.
- Greenhouse gases alter the flow of energy in the atmosphere.
- Water vapor, carbon dioxide and methane are critical greenhouse gases.
- Carbon and water flow through the components of the earth system.
- The climate system has cycles that evolve on many timescales from seconds to millions of years.

Chapter 3
Climate Change and Global Warming

We have discussed the pieces of the climate system and how they interact through some of the key cycles between parts of the climate system. These cycles include the hydrologic (water) and carbon cycles. In this chapter we describe *why* climate changes, with a brief examination of *how* it changes. Details of potential specific climate changes are discussed in Sect. 3.3. The energy budget of the planet is critical for understanding how the climate system may change over time, because on long timescales, climate is governed largely by the total amount of energy in the system and where it goes. Changes to the energy budget, both natural and caused by humans, will cause *climate change*. Because climate is a set of different distributions of weather states in different places, climate change is the altering of some or all of these distributions (e.g., temperature or rainfall at a particular place). *Global warming* implies a specific metric (global average temperature) and a specific direction (warming or a positive trend). Global warming is a subset of climate change.

In this chapter we start with some basic concepts. We start by showing how changes in climate can happen internally in the climate system as a result of coupling of different processes, and we introduce the concept of a feedback (see box) that alters the response of the climate system in reaction to a change to energy that forces the system. We then talk about how the climate system responds to being "pushed": generally with a change in the external heat added or removed. Greenhouse gases trap more heat in the system; hence, they provide this push, or "forcing." We can see how this has happened over the distant (geologic) past and what is currently happening based on the recent (observed) past. Finally, we investigate how the system responds to changes in the energy flow and where it might put the heat: how the changes to the heat input and output may result in climate changes. These are all basic background points for understanding the underlying premise of climate change and, hence, the goals of a climate model to predict climate changes.

© The Author(s) 2016
A. Gettelman and R.B. Rood, *Demystifying Climate Models*,
Earth Systems Data and Models 2, DOI 10.1007/978-3-662-48959-8_3

3.1 Coupling of the Pieces

The different components and processes in the climate system are "coupled" together like a complex and three-dimensional jigsaw puzzle. In this context, coupling refers to the two-way interaction between different parts of the climate system. Take, for example, processes regulating water in the hydrologic cycle. Water evaporates from the ocean, leaving salt and changing density, moving through the atmosphere and depositing heat when it condenses and then precipitates onto the terrestrial surface. All these steps couple water and energy together.

The complex **coupling** of the earth system means that it is constantly evolving on many timescales from a day up to millions of years. The daily timescale is driven by earth's rotation. The annual cycle is the earth's orbit around the sun. There may be small fluctuations in the sun over an 11-year solar cycle, or over different solar cycles. The earth's orbit and wobble on its axis takes thousands of years, and over millions of years the continents also rearrange themselves, affecting the components of the system such as the ocean circulation and ice sheets. And of course there are other events that are not cycles: from volcanic eruptions that put gases into the atmosphere and rearrange the surface of the earth, to meteor impacts such as the impact that likely caused the mass extinction when the dinosaurs died out. These events are external climate **forcing**: They exert an external push on the climate system, usually in terms of a change in energy in the system. When we speak of **natural forcing** of the system, it is the changes on these different timescales and how the components of the earth system interact that govern the evolution of climate.

Focus for a moment on the interactions of the various parts of the earth's climate system. Energy comes in through the sun, modulated by the atmosphere, by clouds, and by hitting the surface of the earth. The energy flows through the system (into the land, snow, oceans, and biosphere), and these components respond. The response (change in energy flow) usually has impacts on other parts of the climate system. This we call a **feedback**. (see box).

Feedbacks

Broadly, a feedback occurs when the input is modified by the output of a process. If you have taken a microphone too close to a speaker, you have experienced a **positive feedback**: Sound goes into the microphone, is amplified by the speaker, comes back into the microphone again, is amplified again, and SCREECH! That's a positive feedback loop. In terms of climate science, a feedback is an internal reaction or response of the climate system to external changes (forcing) that results in more changes to the system (enhanced or reduced forcing).

A positive feedback amplifies a signal. That can mean that changes get larger in either direction. With climate, that means a positive feedback amplifies a change regardless of direction. Think of the example of a snow-covered surface. Being white, it reflects away most of the sunlight. If it warms up and melts, the darker surface beneath absorbs more energy and warms more. This is a positive feedback. It tends to reinforce a small change: Warming melts snow, which melts more snow and causes more warming. But, if there is a cooling that creates more snow, this causes more cooling and more snow. So a positive feedback pushes the system away from its original state: It is destabilizing. This is critical to understand. A positive feedback enhances changes in the direction they start: It enhances warming, but it also enhances cooling. One analogy is the effect of gravity at the top of a hill. If you push a ball forward from the top of a hill, it rolls forward down the hill; if you push a ball backward from the top, it rolls backward down the hill. This is an "unstable" situation, and gravity will accelerate any motion downward: it acts in the same way as the motion.

There are also feedbacks that act to resist changes. This is called a **negative feedback**. A negative feedback tends to push the system back to its original state: a stabilizing force. Note that the connotation is the opposite of typical usage in the context of climate change: negative feedbacks stabilize the climate system (which is usually a good thing), whereas positive feedbacks destabilize it. An example of a negative feedback is the "temperature" feedback: A warmer planet radiates more energy to space, which will reduce the tendency of the surface to warm. To continue the analogy with rolling objects, if you are at the bottom of a valley and push a ball forward (uphill), it rolls backward (down) to where it started. If you push a ball backward (also uphill), it rolls forward to where it started. In this situation of a "valley," gravity acts as a negative feedback or stabilizing force: It acts against the motion. We discuss these feedbacks when we discuss the atmosphere in Chap. 5.

Feedbacks are also important in understanding the climate system.[1] Feedbacks govern the sensitivity of climate to changes (also called climate sensitivity). If we change CO_2 (or more specifically the overall radiative forcing from CO_2 and other gases), how much energy will remain in the earth system? Feedbacks "amplify" (or dampen) radiative forcing, resulting in more or less forcing of the system. If CO_2 increases the energy in the system (and the temperature), this might melt more snow and the darker surface increases the energy in the system by decreasing the albedo (more absorption). This is a positive feedback, which increases the sensitivity of the climate system to changes in the **energy budget**, the amount of energy in the earth system. Changes to the energy budget are of fundamental importance because the overall "level" of energy in the system governs the expected average temperature at the surface. An overall energy change implies, "shifts" in the distribution function of climate (refer back to Fig. 1.3), and the sensitivity tells us the degree of shift for a

---

[1]See Stocker T. S., et al. (2001). "Physical Climate Processes and Feedbacks." In Houghton, J. T., Ding, Y., Griggs, D. J., et al. (eds). *Climate Change 2001: The Scientific Basis. Contribution of Working Group I to the Third Assessment Report of the Intergovernmental Panel on Climate Change.* Cambridge, UK: Cambridge University Press.

given amount of forcing from CO_2 and other gases. So it becomes a useful metric (measure) of where the system might be heading.

We cover feedbacks in more detail when discussing the specific components of the climate system and how we model them in Sect. 3.2. As described in the box, there are both positive feedbacks (like the ice-albedo feedback) and negative feedbacks (like the temperature feedback) in the climate system.

Put a lot of complex feedbacks together, and the climate system starts to sound less like the single screech of an electronic microphone and more like the complex tones of a symphony, perhaps one without a conductor. Many of these tones or combinations of tones have feedback loops, like the example of the sea ice–albedo feedback described here. We discuss this feedback more in Chap. 5, after we introduce additional concepts about how the atmosphere works.

To understand the climate system and how it will evolve over time, it is critical to understand these complex interactions. Most climate science is dedicated to understanding the components of the system, their coupling, and how they work now and have worked in the past, so that we can understand how they might work in the future. The imperative to understand the climate system stems solely from how society is affected by variations in high-frequency extreme weather events and lower frequency climate extremes and how they may change over time in response to different forcings within the system. Or even with no forcing of the system. The importance of understanding weather and climate is independent of any human influence. Even if we did not have reason to think the climate was changing due to human activities, it would still be important and critical to understand and simulate the climate system to predict the natural variability and potential changes in the system so that we can adjust. This is also called adaptation to climate changes, as in adapting our society to a new climate.

3.2   Forcing the Climate System

The **anthroposphere**, the range of human activity, now exerts a strong effect on the climate system. The change in greenhouse gases caused by human activity changes the energy flow in the climate system and creates a forcing on the system. Humans, though largely terrestrial creatures (except for our boats, surfboards, and air travel) are a significant part of the climate system. It is not from the carbon in our bodies. We are small fish, really, compared to all the fish, or rather plankton, in the sea. Our impact comes instead from the carbon from dead plant and animal material that we use as energy. **Fossil fuels** (coal, oil, and gas) are buried sediments of carbon that we pull out of the ground and break apart by oxidizing them at high temperatures. We break carbon bonds in C–H (and C–H–O) compounds, adding oxygen from the air, and get energy, heat, and the chemical by-products: gaseous CO_2 and water (H_2O). There is nitrogen, too, but we consider that later.

Over the last 200 years, accelerating energy use and industrialization has led to the build-up of CO_2 in the atmosphere. We have observed this in a number of ways.

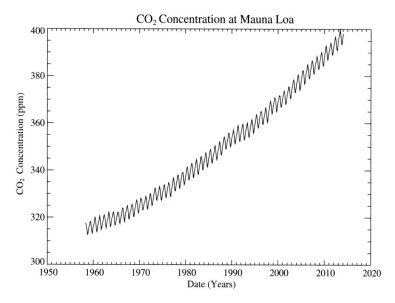

Fig. 3.1 Atmospheric CO_2 concentration from Mauna Loa, Hawaii, in parts per million (ppm) as a function of time

The most direct is through measurements of the concentration of CO_2 in the atmosphere since the 1950s. Every month for 60 years or so, a sample of air is taken and analyzed. Since CO_2 is well mixed in the atmosphere, it represents a broad region (the Northern Hemisphere). Now there are many samples at different stations and multiple instruments, but the answer is the same. The curve in Fig. 3.1 shows two things. One is the upward march of CO_2 concentration over time. The other is the annual cycle of the earth's biosphere: There is an annual cycle in atmospheric CO_2 concentration that occurs because plants in the Northern Hemisphere grow in the spring and turn CO_2 into plant material (leaves, for example), drawing down the concentration. In the autumn, leaves fall and decompose, and much of the carbon returns to the atmosphere. But upward the concentration goes. The curve in the figure is iconic enough to be named the Keeling Curve after the American scientist, Charles Keeling, who first started the measurements in 1958 and made the plot.[2] The units are in parts per million, which means one molecule of CO_2 for every one million molecules of air.

   Some observers like to compare this annual cycle of respiration of the whole biosphere (the sum of life on the planet) to the "breathing" of the planet: equating the earth to a single living being. This is an element of the **Gaia hypothesis** put

---

[2]For a history of the Keeling curve mixed in with the science of climate (and then some policy), see Howe, J. P. (2014). *Behind the Curve: Science and the Politics of Global Warming*. Seattle, WA: University of Washington Press.

forward by James Lovelock and Lynn Margulis in the 1970s.[3] It makes an interesting and powerful analogy that, like a living thing, the earth and its biosphere (the
living sphere) interact to make the whole planet itself seem to act like an organism.
It is also wonderful to note that the biosphere is so entwined with our planet's
climate system that its "breathing" changes the atmosphere. This is another way that
life changes the composition of the atmosphere (the oxygen itself is there because
of life as well). And it shows the importance of life to the cycle of CO_2 in air.

Now let's talk about the increase over time (the trend), the second important part
of Fig. 3.1. There is a steady increase of CO_2 in the atmosphere. This is the result of
human emissions of greenhouse gases. It follows fairly closely with the total
amount of fossil fuel combustion (burning) estimated over the last few hundred
years, starting with the industrial revolution.

We have many lines of evidence that indicate the upward march of CO_2 concentrations is due to human activity. One way is through a direct measurement of
the type of carbon present in the atmosphere as CO_2. Different types of the same
element are called **isotopes**,[4] with different numbers of neutrons in their atomic
nucleus (see box on carbon isotopes). This means they have slightly different mass.
And the relative amounts of the different isotopes of carbon atoms in the atmosphere are looking more like the carbon isotopes in fossil fuels. Plant tissue has a
slightly different balance of carbon isotopes (see box), and the atmosphere is
becoming more abundant in this isotope. This indicates the combustion of dead
plant material from fossil fuels. If the fossil fuels were not causing the increase, we
would not see changes to the carbon isotopes.

Carbon Isotopes
The standard form (isotope) of carbon has 6 protons and 6 neutrons (^{12}C). In
chemical nomenclature, a preceding superscript number on the element
(C) indicates the total number of protons and neutrons. The form resulting
from cosmic rays hitting CO_2 in the atmosphere has 6 protons and 8 neutrons
(^{14}C). ^{14}C is used to carbon date archaeological finds. The atmosphere is
starting to have less of a stable form of carbon that has 6 protons and 7
neutrons (^{13}C). ^{13}C occurs differently in the atmosphere than in plant tissues,
and the proportion of ^{13}C in the atmosphere is of the right proportion as dead
(and fossilized) plant material in fossil fuels.

---

[3]The original paper on the Gaia hypothesis is Lovelock, J. E., & Margulis, L. (1974).
"Atmospheric Homeostasis by and for the Biosphere: The Gaia Hypothesis." *Tellus Series A*
(Stockholm: International Meteorological Institute), 26(1–2): 2–10. There are some good later
books by James Lovelock, including, Lovelock, J. (1988). *The Ages of Gaia: A Biography of Our
Living Earth.* New York: Norton; and Lovelock, J. (2009). *The Vanishing Face of Gaia.* New
York: Basic Books.

[4]For a background on isotopes in the environment, see Michener, R., & Lajtha, K. (2008). *Stable
Isotopes in Ecology and Environmental Science.* New York: Wiley.

3.3   Climate History

In addition to direct observations of the atmosphere for the past 50 years, there exist "fossilized" pockets of the atmosphere: samples of air trapped in ice sheets. Direct measurements of air trapped in ice go back all the way to the formation of the glaciers in Antarctica: nearly 800,000 years of CO_2 measurements. The time is measured first by accumulation of layers of ice, then by dating trace elements that decay in the ice or air bubbles, or counting layers. Figure 3.2 illustrates this record from the Vostok station ice core in Antarctica. Two curves are shown. The first is CO_2 concentrations from bubbles in the ice. Notice how it goes up and down. The times when CO_2 is low correspond to ice ages. At 10,000 years ago, the CO_2 concentration rises and the last ice age ends. Why the cycles? They correspond roughly to some of the changes in the earth's orbit. When conditions favor colder Northern Hemisphere land temperatures, snow sticks around in the summer and ice sheets grow on northern continents. These conditions occur with shifts in the earth's orbit: changes to the tilt of the earth's axis so that a larger tilt gives more severe winters in the Northern Hemisphere, or a shift in orbit so that the earth is farther from the sun during the Northern Hemisphere winter. These cycles are called Milankovitch cycles,[5] after Milutin Milanković, a Serbian geophysicist of the early 20th century. These cycles provide a way of understanding past (and future) variations in the earth's orbit and estimates of the change in solar input (**insolation**) that results. The change in insolation is a natural forcing.

The **ice core record** also contains some interesting signatures in the ice itself.[6] The oxygen in the ice (H_2O) also has isotopes. The forms with more neutrons are "heavier" (^{18}O versus ^{16}O). The heavier form of water tends to remain in liquid phases when water evaporates at the ocean surface, and the relative amounts of heavy and light oxygen in ice thus give a rough measure of the ocean temperature when the ice was deposited. It is not a pure "thermometer" but a relative one, and we can guess at the scale. So oxygen isotopes are used to determine an approximate thermometer and the temperature scale on the right side of Fig. 3.2 is derived from these isotopes (there were no thermometers half a million years ago). The oxygen isotope record is a **proxy** record of temperature.

The story told in the ice is remarkable: The temperature seems to vary in lock step (highly correlated) with the CO_2. When there is more CO_2 (trapping more heat), the temperature is warmer. When there is less CO_2 (less heat trapping), the temperature is colder. We have not said whether the carbon causes the temperature to change or the temperature changes the carbon.

---

[5]For more details on paleo-climate, see Hays, J. D., Imbrie, J., & Shackleton, N. J. (1976). "Variations in the Earth's Orbit: Pacemaker of the Ice Ages," *Science, 194*(4270): 1121–1132.

[6]For background and details of ice core science, a good review is Alley, R. B. (2000). "Ice-Core Evidence of Abrupt Climate Changes." *Proceedings of the National Academy of Sciences, 97*(4): 1331–1334. doi:10.1073/pnas.97.4.1331.

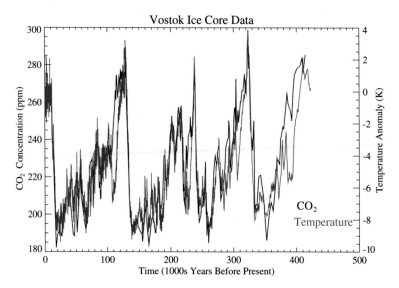

Fig. 3.2 Vostok ice core data. Proxy temperature (from oxygen isotope ratios) in *red* (*right scale*) and CO_2 concentration of air bubbles in *black* (*left scale*). Data from NOAA paleoclimatology program. Petit, J. R., et al. *Vostok Ice Core Data for 420,000 Years*. IGBP PAGES/World Data Center for Paleoclimatology Data Contribution Series #2001-076. NOAA/NGDC Paleoclimatology Program, Boulder, CO

One of the major complexities of the climate system and climate change is that the "problem" or "pollutant" we are discussing (CO_2) is not something "foreign" to the system. It is a part of the system, a critical part that is naturally all around us. We drink CO_2 (in carbonated drinks and beer, for example), we exhale it, and are bodies are made up of carbon. Carbon is absorbed by plants with photosynthesis and used to build their tissues. This creates the natural annual cycle in Fig. 3.1 of carbon in air. So CO_2 is not bad; it is a natural part of the system. The breathing is natural, but the increase is not.

Thus we have direct records of CO_2 in the atmosphere and evidence that recent changes are caused by humans. This is a strong forcing on the system. We also have evidence through proxy records that temperature has been correlated with CO_2. So we know that in the past the earth's climate has changed with CO_2. To link the change in energy (forcing) with the changes in temperature, we need to understand where the energy goes in the climate system. Climate models are one tool for that, but we can discuss the energy flow in more detail.

3.4   Understanding Where the Energy Goes

As discussed in Chap. 2. The earth radiates away energy to space. Greenhouse gases trap some of this energy that would be radiated away. Higher levels of greenhouse gases mean a small fraction of the energy that used to escape stays in the system. Understanding where the energy goes is critical for understanding how the climate might change over time. The energy has to go somewhere. It can start by evaporating water, for example, but this energy will be released as heat when the water condenses. The energy might heat ocean water, and this water might sink away from the surface of the earth (though it is harder for warmer water to sink). It might be radiated back to space somehow. Or it might go to heat up part of the earth system eventually. The challenge is to use what we know about the system to figure out where the energy goes. All of these possibilities imply some change in "climate" somewhere, even if there is no mean temperature change.

The increase in CO_2 in the atmosphere adds more heat to the system. But we are changing the system by only a small amount. Notice that the vertical axis on Fig. 3.1 does not start at zero. Why should that matter? The complexity with understanding climate change is that we are adding a tiny bit of energy to this wonderfully complex system. What is going to happen?

Perhaps the best analogy for adding CO_2 to the climate system is an analogy with our bodies. CO_2 is like a steroid,[7] a natural substance in our bodies that helps our muscles function. Add a bit more, though, and it throws the system out of balance. Steroids enhance muscle performance in the short term (they may also cause long-term damage), enhancing athletic performances. It only takes a small amount of additional steroid to significantly affect athletic performance. The analogy can be taken one step further. How do you know that any single athletic performance was enhanced by an added steroid? A particular basket in basketball, a hit in baseball, a goal in European football (soccer for those in the United States) depends on a lot, not just the steroids in an athlete's body. And the natural steroids vary as well. So it is hard to say that a single home run in baseball, a single goal, or a single time for a distance runner or cyclist, for example, is due to altered performance. But now look at the distribution of that event over time (number of home runs in a season, average time in a race) and the distribution may have changed (more home runs over a season, for example). So it goes with adding carbon to the system and shifting the distribution of climate events (hurricanes are the tropical thunderstorm equivalent of a home run or a goal). This is how we statistically try to ferret out climate change from all the statistical "noise" of weather events.

Returning to the concept of climate as a distribution introduced in Chap. 1, climate change is the change in that distribution. This is illustrated graphically in Fig. 3.3 (a reprint of Fig. 1.3). We often discuss climate as either the average (or **mean**, where the area is equal on either side) or the **mode** (the most frequent occurrence in the distribution).[8] But as we said earlier, no one ever gets killed by the global average temperature. Nor is anyone killed by mean temperature or

---

[7]The analogy between CO_2 and a steroid is usually credited to Jeff Masters and Anthony Broccoli. There is a good video description at https://www2.ucar.edu/atmosnews/attribution/steroids-baseball-climate-change.

[8]Don't be scared by the statistics. See Devore, J. L. (2011). *Probability and Statistics for Engineering and the Sciences*, 8th ed. Duxbury, MA: Duxbury Press, or the terms can be looked up specifically in Wikipedia.

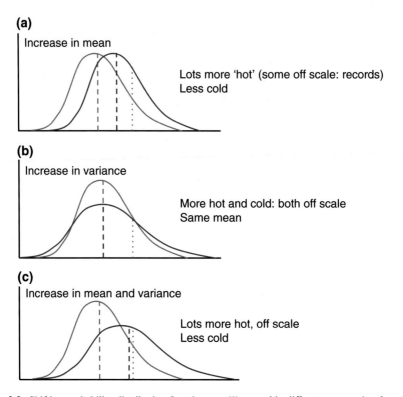

Fig. 3.3 Shifting probability distribution functions are illustrated in different ways going from the *blue* to *red* distribution. The *thick lines* are the distribution, the *thin dashed lines* are the mean of the distributions and the *dotted lines* are fixed points to illustrate probability. Shown is **a** increase in mean, **b** increase in variance (width), **c** increase in mean and variance

precipitation at a given place and time. The extremes (often called the "tails" for their long, skinny graphical appearance) are really the important part. And because they are rare (not very probable; on the graph, a low extent in the vertical), they are hard to predict statistically. Here's a simple example: We often talk about a 50-year flood. This means the flood's "return time" is estimated at 50 years. Or the probability of having such an event in any year is 1/50, or 2 %. If we try to estimate this from a 25- or 50-year record, we may be in error. What if it was a dry period, and in 25 years no floods of a given level were seen? We might conclude that the specific level of flooding can never be higher than what occurred in the last 25 years in a given place. This clearly may not be accurate with a short record. Thus the infrequent tails of the distribution are highly uncertain.

This makes climate change more difficult to estimate. Let's shift the distribution now and assume the climate changes. We can do this first by leaving the shape the same (Fig. 3.3a). Notice what happens to the extremes. At the warm end, the area under the curve beyond some threshold becomes much larger. The area is related to the probability of an event: the fraction of the area is a percent chance of occurrence. And at the cold end, the probability (area) goes way down (gets smaller), even though the mean of the distribution does not change that much.

Here is another example (Fig. 3.3b). Suppose we change the climate by increasing the variability: making the curve wider. The "mean" stays the same, but now the extremes have higher probability in both directions. Here is an example of climate change, without changing the mean (temperature, for example). Think again about living in such a place. Suddenly there is more hot and cold weather, even if the average is the same. In other words, there's a different climate (e.g., more air conditioners or more snow shovels).

Finally, consider a change to both the mean and the distribution (Fig. 3.3c) at the same time. Now one extreme becomes much more probable at the expense of another.

What are the implications? Where does the heat go? When CO_2 is added, the extra heat is absorbed in the atmosphere initially. Recall that it is the heat radiated away from the earth. It's like another thin blanket is added to the thick blankets of greenhouse gases in the atmosphere. The added blanket absorbs a little more heat, which is radiated down to the oceans and land. So there is more energy available at the land surface. There is also more energy available to the ocean surface.

Oceans can warm, but not all the heat added to the ocean will warm the surface temperature. First, some of the heat in the surface ocean may end up in the deep ocean away from the surface. The oceans have a complex circulation, as we will discover in Chap. 6. Some of the water in the ocean is rapidly carried down into the deep ocean in certain regions. If the water contains more heat, this heat will be put deep into the ocean and will not warm the surface. Second, some energy increases evaporation at the ocean's surface and the warmer atmosphere can hold more water. The increased water in the atmosphere can move more heat around. This may not directly heat the surface locally, but it will move heat in the system.

The impacts of these CO_2 changes thus induce several important feedbacks (see box on feedbacks, earlier in this chapter). The first feedback is from additional water vapor that results from warming temperatures. Since water vapor is also a greenhouse gas, adding a little bit of an extra CO_2 blanket to the atmosphere heats the atmosphere by trapping more heat. This allows more water vapor in the atmosphere, which also traps more heat (positive water vapor feedback).

Second, warming due to CO_2 and water vapor may cause changes to the albedo (whiteness) of the planet. This can happen in two ways. Warming can melt snow and ice, or it can change clouds. Melting of snow and ice results in a darker ocean or land surface than when frozen, so more heat is absorbed (a positive snow-albedo feedback). Changes to clouds alter how energy is absorbed or reflected. Clouds are the largest uncertainty in this picture. Clouds broadly cool the planet (they are white and mostly low), but the changes to clouds may warm (if low clouds decrease and the planet is darker) or cool (if the clouds get more extensive).

The resulting changes in the surface temperature and the distribution of heat may change wind patterns. The energy of the water is deposited in different places and changes clouds. The heat going into ice and snow can cause melting when it gets to the melting point. Melting may significantly change the surface albedo. For both clouds and ice/snow, the contrast in color (white ice and clouds versus darker land or ocean) is also important: It causes more energy absorption.

There are several important feedbacks with the terrestrial surface as well. Since plants are made of carbon, they remove it from the atmosphere. Generally, plants get more efficient at growing with more CO_2, just like animals (including humans) do better with more oxygen. If plants have enough water and nutrients to grow, they should increase their growth with more CO_2 in the atmosphere, removing CO_2 into their tissues. This is a negative feedback: More CO_2 enhances its removal by plants. These feedbacks are treated more fully in Chap. 7.

The main point is that increasing CO_2, even a little, throws the earth's climate system out of balance. We have some idea of how it will adjust: There is more heat trapped in this system. This sets off a particular set of feedbacks. Some of the feedbacks are well understood. Some feedbacks are not well understood. Some feedbacks are positive; some are negative. We do not fully know exactly where all the extra energy will show up: as wind, as heat, or as rain. We expect the distribution of climate to evolve over the planet.

3.5   Summary

This chapter has sketched out the essence of climate change. The climate system as a whole responds to a forcing in complex ways. The complexity arises because the different parts of the climate system are coupled together. There are many feedbacks in the system. We have a good idea of the past and present forcing that is pushing on the climate system. For the recent past, we have strong evidence that this forcing is from a buildup of CO_2 in the atmosphere over the past 150 years. From the composition of the atmosphere (isotopes), we know this is a result of human activities: We are changing the very composition of the atmosphere.

Section III of this book will treat in detail the uncertainties in climate prediction, but we can make some broad statements. When discussing future climate change, we usually mean **anthropogenic** (human-caused) climate change, where the change is in response to a forcing from humans. We also can discuss natural climate change. But the human-caused climate change is because of more CO_2, trapping more energy in the system. Globally, the distribution of global average temperatures is expected to shift toward warmer conditions; hence, we sometimes refer to anthropogenic climate change as "global warming." The change in the regional and local distribution of climate variables (temperature and precipitation) might be expected to increase warm extremes at the expense of cold extremes. But the distribution shape may shift over time in ways we do not yet understand. Since different places have different climates with different distributions, they may change in different ways. This might mean big differences in extreme events (the tail of the distributions in Fig. 3.3): tropical cyclones, extended droughts.

We have theories about how the different feedbacks in the climate system work based on observations from past and present climates. We discuss these feedbacks in the context of models in the next chapters. To confirm our understanding, we also try to use models of the system to estimate what has happened in the past and what will happen in the future. We can do this by applying forcing to models and observing how they respond. Thus climate models are used to translate the basic constraints on climate from forcing and feedbacks into specific predictions about regional or local climate changes.

Key Points

- Understanding how parts of the climate system are coupled with feedbacks is critical.
- Greenhouse gases (CO_2) have been increasing over the past 60 years (based on measurements of air samples) and for the past 150 or so (based on ice cores).
- The composition of the atmosphere tells us that the increased CO_2 comes from fossil fuels.
- Increasing greenhouse gases trap more energy in the system. The energy has to go somewhere.

Chapter 4
Essence of a Climate Model

Now that we have described climate, we ask the question, what is a climate model? A climate model is a set of equations that try to represent and reproduce each of the important pieces of the climate system. The model is developed based on everything we know about the world around us. Traditionally and historically, there are different sections in the model (sub-models, or components) for the major spheres we have discussed: atmosphere, ocean, land, cryosphere and biosphere. If these different spheres can interact with each other in a simulation, then we say they are coupled together.

Although a model can sometimes be a physical object (a model airplane, or a physical model of a building), a climate model exists as a conceptual model coded into a computer (think of the drawings of a building's plans on a computer). The structure of the model is a description of the physical laws of the system. It is a series of equations. These equations are a description of the climate system: component by component (e.g., atmosphere, ocean, land), process by process. The set of equations is analogous to the description of a building contained in blueprints that describes the structure, components, dimensions and finishes. This description can be used to simulate the building in three dimensions so that you can see what the building will look like in the future when it is built. Not unlike a climate model, the structure of a building is also governed by fundamental physical laws: We discuss them in Sect. 4.2. However, climate models are dynamic, meaning they change in time. Although a building may seem static, many complex structures, including buildings, are described and subjected to simulated forces (e.g., to simulate earthquake effects) on a computer to understand how they might react.

The equations in a climate model can be (and are) written down on many pieces of paper; description documents run to hundreds of pages.[1] To solve these equations efficiently, a computer is used. A "simple" climate model can be written out in just a few equations and either solved by hand or put into a spreadsheet program to solve. We illustrate the concepts of such simple models below. More complicated

---

[1]For example, Neale, R. B., Chen, C. C., Gettelman, A., Lauritzen, P. H., Park, S., Williamson, D. L., et al. (2010). *Description of the NCAR Community Atmosphere Model (CAM5.0)*. Boulder, CO: National Center for Atmospheric Research, http://www.cesm.ucar.edu/models/cesm1.0/cam/docs/description/cam5_desc.pdf.

© The Author(s) 2016
A. Gettelman and R.B. Rood, *Demystifying Climate Models*,
Earth Systems Data and Models 2, DOI 10.1007/978-3-662-48959-8_4

models are essentially giant spreadsheets inside of supercomputers. We also discuss how different types of models are constructed. Finally, we discuss exactly what it means to set these models up and "run" them on large (super) computers. These methods give us are a general way to think about climate models before diving into the details of what the models contain.

4.1 Scientific Principles in Climate Models

Each of the components (submodels) and the individual processes must obey the basic physics and chemical laws of the world around us. An important overlooked fact is that the fundamental principles of climate modeling are not new. Simulating the earth system relies on principles of physics and chemistry that have been known for 100–300 years. The existence of a new subatomic particle does not require us to change our climate models. They contain no complex physics (like presumptions of warping space-time).

The physical laws start with **classical physical mechanics**,[2] developed by Sir Isaac Newton in his *Philosophiæ Naturalis Principia Mathematica* (1687): conservation of mass and momentum (especially Newton's second law of momentum) and gravity. The classical physical mechanics of the atmosphere and ocean (air and water) use equations developed by Claude-Louis Navier and Sir George Gabriel Stokes in the first half of the 19th century, known as the Navier-Stokes equations. The same equations are used to simulate airflow around aircraft, for example, in another type of finite element modeling.

In addition to the motion of parts of the earth system, flows of energy are critical in the climate system. As we discussed earlier, the slight imbalance of energy input and outflow as carbon dioxide concentrations increase gives rise to climate change. The transformation of energy and its interaction with the physical system is known as **thermodynamics**,[3] the principles of which were developed by Nicolas Carnot and others in the early 19th century. Flows of energy are essentially **electromagnetic radiation**, described by the electromagnetic theory of James Maxwell in the 1860s. Important details about how radiation interacts with thermodynamics were added by Jozef Stefan and Ludwig Bolzmann in the 1870s and 1880s. Also in the 19th century, much of the basic work on chemistry was performed, culminating in

---

[2]Starting with Newton, there are many books on the subject. Perhaps the best modern reference is still the *Feynman Lectures on Physics*. You can buy them, but they are available online from http://www.feynmanlectures.caltech.edu/. Classical mechanics is Volume 1, mostly Chaps. 1–10.

[3]*Feynman Lectures on Physics* (http://www.feynmanlectures.caltech.edu/), Volume 1, Chaps. 44–45. Or there is always Pauken, M. (2011). *Thermodynamics for Dummies*. New York: Dummies Press.

specific experiments and estimates by the Swedish chemist Svante Ahrrenius in the late 19th century about the radiative properties of carbon dioxide.[4]

In the face of criticism of climate science, it is important to note that the physical science behind climate models and energy is based on physical laws known for several hundred years and is not new or subject to question. If the world did not work this way, cars would not run, airplanes would not fly, and everyday motions that we observe (baseball pitches, gravity) would not happen. As we demonstrate later, these underlying scientific principles are not cutting-edge science. The principles are not open to question or debate, any more than the law of gravity can be debated.

Climate models simply take these basic laws, apply them to a gridded representation of the different pieces of the earth system and connect it all together. The overall philosophy is classic scientific reductionism. The same principles and scientific laws are used in countless other fields. Do we "believe" in climate models? That is a bit like asking if we "believe" that the earth is round, that the sun will rise in the east, or that an airplane will take off when it gets to a certain speed. But if you still don't, please reread the "Models All Around Us" box in Chap. 1. We use physical laws that agree with observed experience to make a prediction. This is a different way to use models than many people are used to (see box on dynamical system models below).

Dynamical System versus Empirical Models
Weather and climate are dynamical systems; that is, they evolve over time. Dynamical system models use equations of relationships between variables to describe the future state of a model. The future state of a dynamical system is dependent on the present state. Scientists in many fields use models that describe dynamical systems with time evolution.

The rules that define the evolution of the Earth's climate rely on physical laws and relationships. So, for example, the speed or velocity (v) of air is defined by the equation that describes the conservation of momentum. The velocity of a "parcel" of air is the existing velocity (v_0) plus the acceleration of the object (a) over a given time interval (t). So $v = v_0 + at$. This is based on Newtonian mechanics, the basic laws of common physics. If the desired output is the velocity v at any time, then the inputs are v_0, a, and t. The equation can be marched forward in time (where at the next time $v_0 = v$ from the previous time). This equation predicts how the state (physical properties: velocity, in this case) of the object changes over time. Climate models have equations of motion for air, water, ice and the biosphere that are integrated forward in time.

A different way to represent a dynamical system is with a statistical or empirical model. Empirical models define mathematical relationships

---

[4]The original paper: Arrhenius, S. (1896). "XXXI. On the Influence of Carbonic Acid in the Air Upon the Temperature of the Ground." *London, Edinburgh, and Dublin Philosophical Magazine and Journal of Science*, *41*(251): 237–276.

between independent variables (inputs) and dependent variables (outputs). For example, if you measure the speed of an object at different points in time, you can develop a relationship based on those observations. If you are dropping the object with no air resistance, so the acceleration is gravity, you can develop a relationship between the velocity and the time. For the surface of the earth, you would get (in metric units) $v = v_0 + 9.8\,t$, where t is measured in seconds, and v is in meters per second. This is an approximate form of the equation of motion, which might work very well for similar cases, but would not work for a different situation.

Physical laws contain more information than statistical or empirical methods and, therefore, are more suitable for dynamical systems where the environment for statistically based parameters might be different. For example, the gravitational acceleration is dependent on the mass of the object that is doing the attraction and the distance from the center of that mass (Earth, in this case). So the dynamical system approach works on the moon: You can calculate different acceleration (a) based on the lunar mass. But the empirical result (using 9.8) would not work on the moon.

The danger with statistical or empirical models is being "out of sample": There is some condition where the model does not work. This may be obvious in our example, but it is not always obvious.

So are dynamical models always better? Only when a good description of the system can be made. For many processes, we turn to empirical or statistical relationships. Even many fundamental properties of the world around us are made up of many different conditions at the molecular or atomic level, so we have to describe the process empirically. As an example, the chemical properties of a substance, like the freezing temperature and pressure of water, are related to small-scale motions of molecules (all governed by our velocity equation), but we cannot measure each molecule. So we measure the collected behavior of all the molecules in a sample and build an empirical model of the freezing point of water as a function of temperature and pressure.

Thus climate models do contain empirical models of processes, coupled together in a dynamical system. They contain a representation of the freezing point of water, for example. These processes are tied together using physical laws, which help us to make sense of the interconnection between the processes. Some processes are simple or well described (like water freezing), and some are very complex. But these statistical models are sometimes necessary. Tying them together with physical laws (like conservation of energy and mass) is an important constraint on climate models. These conservation constraints help to reduce uncertainty.

4.2   Basic Formulation and Constraints

Ultimately a climate model is a series of interlinked *processes* and a set of equations or relations: physical laws that control how the system evolves. These different laws are solved for each different location in the model: a finite element. Let's describe how we break up a model into different pieces, what each of these pieces does and why. This will define the basic formulation of a climate model.

4.2.1   Finite Pieces

The physical laws (see below) are solved at each physical location (**point, cell,** or **grid box**) defined in a model. The physical points are illustrated in Fig. 4.1. Most models also have a vertical dimension (whether in the atmosphere, the ocean, or through the thickness of sea ice or soil), making a **column**. This has one dimension: in the vertical. Columns are generally on a regular **grid**, so each location is a grid point. *Grid* comes from a regular lattice of points, usually equally spaced, but they can be irregular (different arrangements of points), which we discuss later. Thus, a model (like the reality it represents) has three dimensions: one horizontal and two vertical (Fig. 4.1c). Each individual vertical location in a column is called a grid box, or cell. Each of these cells is a "finite element" for which a model defines different processes, usually representing a given region with a single "finite" value.

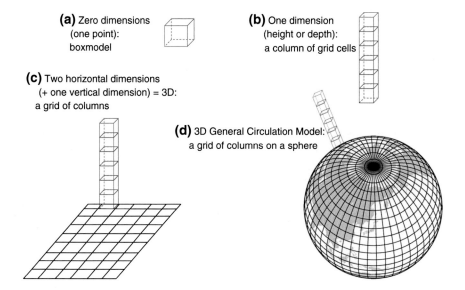

Fig. 4.1 Dimensions of models and grids. **a** Point or box model (no dimensions). **b** Single column (one dimension in the *vertical*). **c** Three dimensional (3D) model with *two horizontal* dimensions and *one vertical* dimension. **d** 3D grid on a sphere

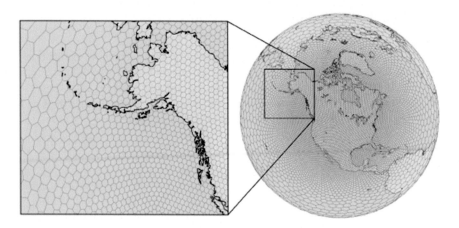

Fig. 4.2 An example of a variable resolution grid from the model for prediction across scales (*MPAS*). The grid gets finer over the continental United States using a grid made up of hexagons. *Source* http://earthsystemcog.org/projects/dcmip-2012/mpas

When we talk of the **resolution** of a model, we mean the size of the horizontal boxes or, equivalently, the space between the centers of different grid boxes. So a model with horizontal resolution of one degree of latitude has grid boxes that are 68 miles (110 km) on a side.

The global grid in Fig. 4.1d is along latitude and longitude lines. It has the same number of boxes in longitude (around the circle) at any latitude. Since the circumference of the earth is smaller at higher latitudes, the grid has unequal areas. This is a problem for several types of model (see Chap. 5, on the atmosphere, and Chap. 6, on the ocean). Some models use other grids to make the different boxes have nearly equal area (e.g., a grid of mostly hexagons). Other grids are designed with higher resolution (smaller size grid cells) in a particular region. This provides benefits of a higher resolution model, but with lower computational cost. Figure 4.2 shows an example of a variable resolution grid.

Motions in the climate system are both horizontal and vertical. Climate models need to represent processes in both directions. Horizontal processes include the flow of rivers, wind-driven forces on the ocean surface, or the horizontal motion of weather systems in the atmosphere. Many features of the climate system vary in the horizontal. The ocean surface is pretty uniform, but the terrestrial surface is not: Vegetation and elevation change. Figure 4.3 illustrates horizontal grids in a climate model, illustrating with horizontal resolutions of about 2° of longitude (124 miles, or 200 km), ~1°, ~0.5°, and ~0.25° (the latter is 16 miles, or 25 km). The color indicates the elevation, showing that, as the resolution gets finer, more realistic features (like the Central Valley of California) can be resolved. For the terrestrial surface, this also means that the land surface (soil, vegetation) can also vary on smaller scales.

There are also many vertical processes: like the rising or sinking of water in the ocean, the movement of water through soil, or the vertical motion of air in a

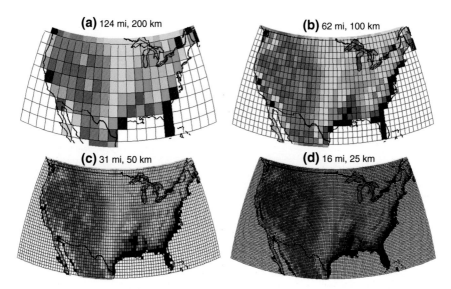

Fig. 4.3 Example of a model with different horizontal resolutions on a latitude and longitude grid over the continental United States. Resolutions are **a** 2° latitude, **b** 1° latitude, **c** 0.5° latitude, and **d** 0.25° latitude. Elevation shown as a color

thunderstorm. These vertical processes feel the effects of gravity and the effects of buoyancy. Buoyant objects are less dense than their surroundings (air or water) and tend to rise. There are also forces like pressure that act both vertically (pressure decreases as you get farther from the bottom of the atmosphere or ocean) and horizontally (wind tends to blow from high to low pressure).

4.2.2   Processes

It seems natural to be able to break down the problem into a series of boxes in physical space for each component, as in Fig. 4.1. But what is in these boxes? Each box tracks the properties of the physical **state** of the system: a collection of variables representing the important physical conditions at a location and time. These are the physical properties and energy in the box: like the temperature of the air in the box. The physical properties include the mass of water or ozone molecules in a box of air, the salt in a box of ocean, the soil moisture and vegetation cover of a box on the land surface. The "state" also records the total energy in a grid box. The total energy has several parts, including the **kinetic energy** (winds, currents, stream flow) of the air or water or ice in motion, and the **thermal energy**, usually represented by temperature. Each of these quantities can be represented by a number for the box: the number of molecules, the temperature, the wind speed, and the wind direction. This set of numbers is the state of box. Figure 4.4 indicates how these

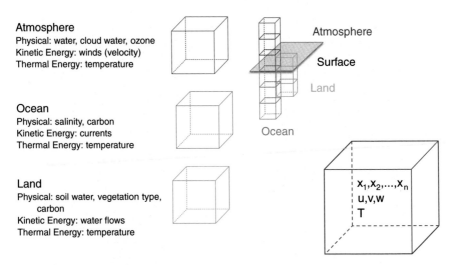

Fig. 4.4 State of the system. Different grid columns for the atmosphere (*red*), ocean (*blue*) and land (*green*) with description of contents. Also a grid box (*purple*) with a 'state' vector of temperature (T), wind in 3 dimensions (U, V for *horizontal* wind and W for *vertical* wind) and the mass fraction of compounds like water (X_n)

components are described by a series of numbers grouped into physical quantities (x_1, x_2, ... x_n for n tracers like water), kinetic energy (u, v, w for wind vectors in three dimensions), and thermal energy (T for temperature).

The basic goal of a climate model is to take these physical quantities (the state) at any one time in each and every grid cell and then to figure out the processes and physical laws that will change these quantities over a given time interval (called a **time step**) to arrive at a new state in every grid cell. Figure 4.5 illustrates the different parts of a time step in a model. A time step involves several different processes. (1) Calculating the rates of processes that change the different quantities with sources and loss of energy or mass and their rearrangement. (2–3) Estimating the interactions between all the boxes (2) in one model column and (3) between different component models. (4) Solving physical laws that govern the evolution of the energy and mass. (5) Solving physical laws for motions of air and everything moving with the air on a rotating planet.

First, processes that change the state of the system are calculated. This might include, for example, the condensation of water into clouds, or freezing of ocean water to sea ice. This is illustrated in (1) in Fig. 4.5. As part of this endeavor, exchanges between boxes in a column are often calculated (2). These steps define the sources and loss terms for the different parts of the state: the quantity of water precipitating, or the quantity of salt expelled by newly formed sea ice. These terms are used in (3) to exchange substances with different components: for example, precipitation hitting the land surface. Then, all these terms for the mass changes in substances and the energy changes are added to the basic equations of thermal

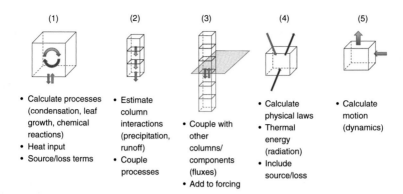

Fig. 4.5 Changing the state: one time step. Climate model calculations in a time step that change the state of a model. *1* calculate processes, *2* estimate column interactions like precipitation, *3* couple with other columns and components, *4* calculate physical laws like radiation, *5* estimate motions

energy (4). Finally, these terms are used as inputs to the equations of motion to calculate the changes to wind and temperature (5).

The physical laws are the fundamental **constraints** on the model state. The constraints are rules set in the model that cannot be violated. In classical physics, mass is not created or destroyed. If you start with a given number of molecules of water, they all have to be accounted for. This is called **conservation of mass**. There is usually an equation for each substance (like water). It has terms for the motion of water in and out of a box and for the processes that transform water (sources and loss terms).

Energy is also conserved. There are equations for the kinetic energy (motion) and for the thermal energy (temperature). There is also potential energy (work against gravity). The total conserved energy includes all these kinds of energy. There can be transformations of this energy that seem to make it go away: Heat is needed to evaporate water and change it from a liquid to a gas. The heat energy becomes part of the chemical energy of the substance. Temperature or heat energy is the kinetic energy of the molecules of a substance moving around, so evaporating water into vapor adds energy to the water, which must come from somewhere. The heat is released when the water condenses to liquid again. This is **evaporative cooling** when you evaporate liquid (and **latent heating** when it condenses). It is also what happens when you compress air (it gets warmer). But the processes will reverse their energy, conserving it. These constraints are quite strict for climate models: If you start with a certain amount of air or water, it has to go somewhere. The transformations and transport (motion) of mass and energy must be accounted for. These properties of physics (at the temperatures and pressures of the earth's atmosphere and climate system) do not change, and these laws cannot be repealed.

Finally, all of these terms are balanced between the grid boxes. This is illustrated in step (4) in Fig. 4.5. Typically, models are used to calculate how the system would change due to different effects. Then these effects are added up. For example, the land surface has water evaporate from it and plants that take up water. The amount of water in the soil column (a box of the land surface) is a result of precipitation falling from the atmosphere, runoff at the surface, and the motion of water in the soil column. These boxes also then exchange their properties with other boxes, such as water filtering deeper into the soil, or runoff going to an adjacent piece of the land surface. In addition, exchanges can occur with other pieces of the system: The precipitation falls onto the land from the atmosphere, evaporation goes into the atmosphere, and runoff goes into the ocean. These interactions can all be described at a particular time, and the effects can be calculated and used to update the state of the system.

Key to this system are the descriptions of each process. Some examples of *processes* are the condensation of water vapor to form clouds, carbon dioxide uptake by plants, or the force on the ocean from the near-surface wind. Each of these processes introduces a forcing on the climate system. As we will learn, many of these processes are hard to describe completely, particularly for processes that occur at scales much less than the typical size of one grid box in a model. A climate model usually has one value for each substance (like water) or the wind speed in each large location, and it has to represent some average of the process, often by approximating key parameters.

Parameterization is a concept used in many aspects of climate models (see box). The basic concept is like that of modeling itself: to represent a process as well as we can by approximations that flow from physical laws. Many of the approximations are required because of the small-scale nature of the processes. The goal of a parameterization is not to represent the process exactly. Instead, it is to represent the *effect* of that process at the grid scale of the model: to generate the appropriate forcing terms for the rest of the system and the rest of the processes.

Parameterization

Representing complex physical processes (clouds, chemistry, trees) in large-scale models is in some sense impossible. The French mathematician Laplace articulated a thesis of the reductionist worldview in the early 19th century: If one could have complete knowledge of every particle in the universe and the laws governing them, the future could simply be calculated. Of course, we cannot do that, so we seek to represent what we know about the behavior of particles, based on physical laws and empirical observation. For some processes, we can refer to the basic physical laws, which often have little uncertainty in them. The laws of how photons from the sun move through a well-mixed gas such as air are an example. Other processes are more complex, or variable on small scales. It is hard to derive laws from these processes. For instance, the flow of low-energy photons from the earth through air is somewhat uncertain because the laws governing how the energy interacts with water vapor are very complex. In the case of water vapor, the way the molecule is constructed it can absorb and release energy at many different wavelengths. For these processes, we often must use statistical treatments to match observations to functions that can be used to describe the behavior. Some processes can be represented by basic laws, other processes must (at the scale of a climate model) be represented by statistical relationships that are only as good as our observations (see box on dynamical vs. empirical models earlier in this chapter).

What processes to represent and at what level of detail are other critical choices. Herein lie decisions that require a rigorous attention to the scientific method: Hypotheses must be developed and tested against observations to ensure the results of the parameterization match observations of the process being represented. To some extent, the complexity may be dictated by the available inputs: If the inputs are only crude and broad in scale, or uncertain, then it may not make sense to have complex processes acting on bad inputs if simpler solutions are possible. But if a lot of information is available, it should be used.

Another determinant is how important the process is to the desired result. Climate modelers worry quite a bit about having detailed descriptions of the flows of energy and mass, especially of water mass (which, by carrying latent heat, affects both energy and mass). Small errors in these terms over time might result in large biases (if energy is "leaking" from the system). So conservation is enforced. But this does not constrain important effects. For example, while total precipitation might be constrained, some of the details of precipitation, such as timing and intensity, are not well represented. Weather models, however, focus much more on the timing and intensity of precipitation by having more detailed descriptions of cloud drops and their interactions, but they often do not conserve energy and mass perfectly over the short period of a forecast.

Putting the processes together seems like a daunting task. It would also seem that one simply is multiplying uncertainty by taking one uncertain process after another. But in fact the physical laws are strong overall constraints on climate models. If each process is bounded and forced to be physically reasonable— starting with the conservation of energy and mass, but usually extending to other fundamental observations—then it is expected the whole climate system being simulated will be constrained but still have the interconnectedness needed to generate the complex and chaotic couplings that we see in the real world around us. The danger is that the complexity gets large enough that we cannot understand it in the model. The rationale is that by interlocking the carefully designed and constrained parameterizations in a sensible way, like putting bricks together, we can build the **emergent** whole of the climate system. The whole "emerges" from a series of processes tied together.

The emergent constraints arising from conservation is where "art" seems to come into climate modeling. How can a crude representation of processes possibly represent the complexity of climate? The constraints drive simulations toward reality. Hence, climate modeling is often called an art, but in a derogatory way, to imply that it does not follow the scientific method. But parameterization development is a series of hypothesis-testing exercises, forming and testing hypotheses for representing processes in the earth system and the connecting of the processes together. The problem is that our incomplete knowledge and imperfect observations permit multiple states of the system that behave similarly: More than one description may match current observations of the earth.

Consider this simple example: Viewed from a satellite in space, the Arctic Ocean appears white. But is it covered by clouds, or just by sea ice? Either option would fit the observation, as both clouds and sea ice are white. But clouds in the atmosphere and ice at the surface are very different and will respond differently to changes in winds and temperatures. If you assume that the "average" condition, or the distribution of how often clouds and ice are present, is not known, then we cannot determine the present climate state from observations. In this case, very different climates with different clouds and ice are possible. The different climates may respond to climate changes in different ways.

The goal of modeling is to try to reduce these uncertainties by careful application of numerical tools to represent climate processes and continual testing against observations. Multiple models and multiple approaches in the global scientific enterprise are competing in this context to see which representations seem to work the best.

It all comes down to representing processes.

But the compensation in a climate model is the conservation laws: Energy and mass must be conserved. Each process at each time step must be limited to what is possible. For example, the amount of water that can rain out of a cloud is limited by the total water in the cloud.

The respect for these fundamental laws and the equations that describe the motion provide strong constraints. If each process is limited and each set of processes in the atmosphere and ocean are limited, then the emergent whole of the sum of those processes is constrained by known physical laws. The complex interactions are constrained by those laws. The model cannot go "out of bounds" for any process, or for the sum of any processes at any time step. This requires the model to be "realistic": resembling the laws of the physical world and the observations of the world. There is no guarantee or theory that prescribes this at the scale of a climate model yet, but energy and mass conservation are powerful constraints.

As we shall discover, there are many different possible representations of processes and their connections in the climate system that are physically realistic. We do not understand the whole climate system well enough to make unique models of each process: Multiple different models are possible. This yields multiple ways to develop and construct a climate model. Different representations will yield different results, sometimes importantly different results. But it also means a "hierarchy" of models is possible: from simple models that try to simulate just the global average temperature, to detailed regional models that try to represent individual processes correctly. These different models are used for understanding different parts of the climate system.

4.2.3 Marching Forward in Time

The reductionist approach to individual effects or processes and discrete time steps is an important part of understanding finite element models such as climate models. For climate models, many decisions can be made, starting with which processes to include and how to represent them.

Figure 4.6 illustrates one method of taking the different physical processes and equations in Fig. 4.5 and marching forward in time. It is drawn as a loop, because where one time step ends another begins. Here the processes and exchanges are highlighted. They occur at every point in every column on the grid for each component model. First shown are (1) the physical (including chemical) processes in each grid box. These interact in the column (2) for example: precipitation falling. There are (3) exchanges between components–like precipitation hitting the surface. Then there is the application of the physical laws. Conservation of mass is applied throughout. Conservation of energy happens in the thermodynamic equation when

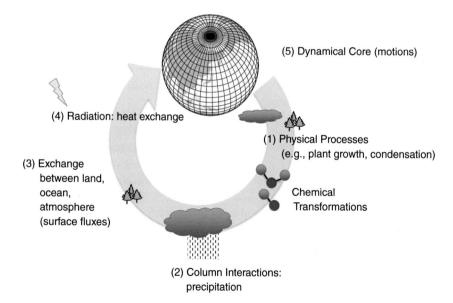

Fig. 4.6 Marching forward in time within a climate model. Time step loop typical of a climate model. Processes are calculated in a sequence at each time. *1* Physical processes and chemical transformations, *2* column interactions, *3* exchange between different components, *4* radiation and heat exchange, *5* dynamics and motion

radiation is calculated (4). For models with a moving fluid or solid like the atmosphere, ocean and sea ice, the equations describing motion (kinetic energy) are solved in the *dynamical core* (5) of the model. The dynamical core solves the equations of motion to determine how substances (water, air, ice, chemicals) move between columns.

But decisions involving which processes to calculate first and how to put them all together, are not always obvious. This is one of the inherent complexities in finite element models. Some choices matter for the results and a lot of research has gone into understanding these choices and the range of solutions that result. One goal is to develop formulations so that the solutions do not depend on the ordering of processes. Fortunately, many of the basic scientific principles used limit the realistic choices, as we have already seen above with the conservation of energy and mass.

4.2.4 Examples of Finite Element Models

Global climate models are made up of a series of component models (e.g., atmosphere, ocean, land). Each component model has a series of grid boxes or cells, on a regular grid. The solution of all the processes and transformations is carried out at each time step, for each one of these finite elements (grid cells) in the model. The concept of a finite element model is used in many other scientific and engineering endeavors. The flow of air over the wing of an airplane is a close analog of many of the concepts used in modeling the atmospheric part of the climate system. Fluid flow in a pipe is another example of finite element modeling. Such models are used for a water treatment plant, a chemical plant, an oil refinery, or the boiler in a coal-fired power plant. Finite element models are also used to understand how engines work in cars and trucks, or how materials perform under different forces (stress), whether an individual part of a device, or an entire structure (a building, an engine block, etc.). These models are used all the time in engineering things in the world around us. The fact that planes fly, cars run, and all our electronic and mechanical devices work is testament to the power of finite element modeling. It includes whatever electronic machine you are reading this on, or whatever machine printed the words in ink on the page you are reading. Numerical modeling works in many fields, and includes many of the same scientific concepts, as in climate modeling.

4.3 Coupled Models

Currently, all the components of the climate system have also been included in earth system models. Generally, the process started with representing the atmosphere (see below for more discussion of model evolution). Representations of the land surface and the ocean were added next and then coupled to the atmospheric model to make a **coupled climate system model**. A climate system model does not include a comprehensive and changeable set of living components in the biosphere. The biosphere contains the flows of carbon in land-based plants and in small organisms in the ocean (phytoplankton). Including the biosphere allows these stocks of carbon to affect the carbon dioxide in the atmosphere. This is a more complete description of the system usually termed **earth system models**. We refer to these models simply as climate models. So where the scientific literature says "earth system," we use the term "climate system."

The components of the earth's climate system (atmosphere, ocean, land) are each coupled to the others as physically appropriate. Figure 4.7 shows a simple schematic of the arrangement. The bottom of the atmospheric model is the top of the ocean and land models, for example. Information is exchanged across the components at these natural boundaries. The exchanges are critical to the operation of the system. Rain falling out of the atmosphere is critical for the state of the land (determining soil moisture, runoff, plant growth, and the like). The interaction of floating sea ice with the surface ocean is critical for the density of the ocean at high latitudes (since when ice freezes the salt is expelled and the water becomes saltier and denser). And the atmospheric winds drive ocean currents and move the sea ice around.

Many of these interactions are illustrated in Fig. 4.7. These interactions are critical for understanding how the climate system evolves and how it responds to changes in the interactions. Small changes in one component have ripple effects on other components. Note how the arrows in Fig. 4.7 can circle back: changing temperatures can melt sea ice. The melting sea ice exposes darker ocean. The darker ocean absorbs more energy. The absorbed energy changes temperature further. These are expressions of feedbacks in the system (see Chap. 5). Each of these component models contains a certain amount of complexity related to the respective piece of the system that a given model illustrates. We address those complexities in later chapters. In this chapter, we are concerned with understanding the essence of these models.

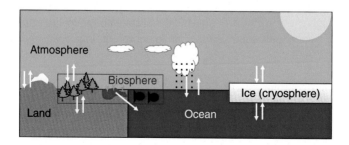

Fig. 4.7 Schematic of earth system coupling. The basic coupling between different components of a climate model

4.4   A Brief History of Climate Models

So how did we get to breaking up the planet into tiny boxes on massive numerical computers humming away in air-conditioned rooms? Climate predication is an outgrowth of wanting to know more about the fundamental and long-term implications of daily weather phenomena.[5] It arose in the 1960s in parallel with the development of weather prediction. Weather prediction actually started well before electronic computers. As mentioned earlier, in the 19th century, scientists speaking as philosophers, such as Simone Laplace, articulated the idea that if we knew where every particle in the universe was and we knew the laws governing them, we could calculate the future. That remains a philosophical statement more than anything else, especially since quantum physics has shown that you cannot measure the characteristics of a particle without affecting them (Heisenberg's uncertainty principle).

Early experiments with forecasting the weather, for example, by Vilhelm Bjerknes in the early 20th century, articulated that with a sufficiently accurate (but not perfect) knowledge of the basic state and reasonably accurate (but not perfect) knowledge or approximations of the laws of the system, prediction was possible for some time in the future. During World War I, a British scientist (working as an ambulance driver) named Lewis Richardson attempted to write down the laws of motion and, using a series of weather stations, calculated the future evolution of surface pressure. These equations were correct but virtually impossible to solve practically by hand. Approximations for the equations, developed by Carl Gustav Rossby in the 1920s and 1930s, enabled some measure of the evolution of the system to be described and enabled some rudimentary attempts at predicting the evolution of weather systems. Electronic computers were developed during and after World War II. One of the first was developed to calculate the tables for the trajectory of artillery shells. After the war, other computers were applied to solve Rossby's simpler set of equations, among others by a group at Princeton led by John von Neumann.[6]

The use of electronic computers to solve the equations of motion describing weather systems led to actual numerical forecasts. So where does climate prediction come in? In the mid-1950s, several experiments took rudimentary weather forecasts, added some of the forcing terms for energy and radiative transfer, and tried to run them to achieve some sort of statistical steady-state independent of the initial conditions. These experiments were able to represent important aspects of the

---

[5]A good overview of the co-evolution of weather and climate models is contained in: Edwards, P. N. (2010). *A Vast Machine: Computer Models, Climate Data, and the Politics of Global Warming*. Cambridge, MA: MIT Press. Another good reference is the description of the history of General Circulation Models in Spencer Weart' online book *The discovery of Global Warming*, Harvard University Press, 2008. Available at: https://www.aip.org/history/climate/GCM.htm.

[6]For a detailed description of the origin of digital computers, focused on von Neumann and the Princeton group (with cameo appearances by climate models), see Dyson, G. (2012). *Turing's Cathedral: The Origins of the Digital Universe*. New York: Vintage.

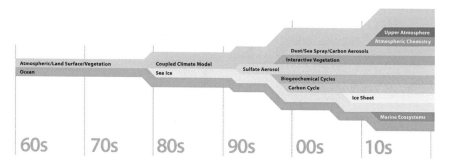

Fig. 4.8 Schematic of components. Evolution of the parts of the earth system treated in climate models over time. *Source* Figure courtesy of UCAR

general circulation, and from there, climate models (originally General Circulation Models) were born. Initial development as a separate discipline evolved in the 1960s. As computers got faster and techniques got better, more realistic simulations evolved. Since then, there has been a co-evolution of climate and weather models with improved computational power.

This co-evolution led to the expansion of climate models from just models of the atmosphere, to coupled models of the entire climate system. Figure 4.8 illustrates how climate models have evolved from simple atmospheric models and ocean models, to coupled models with land and sea ice by the 1980s, to adding particulates and chemistry in the atmosphere, dynamical vegetation and chemical cycles on land, and marine ecosystems and climate in the early 21st century.

4.5   Computational Aspects of Climate Modeling

Climate models are naturally computer codes. They are run on supercomputers. What does it actually mean to *run* a climate model code? What does it entail?

4.5.1   The Computer Program

A climate model is a computer program. Generally each component, such as the atmosphere, can be run as a separate model, or coupled to other components: often a coupled climate model. Figure 4.9 illustrates a schematic of a coupled climate model. The figure is really an abstraction from Fig. 4.7: without the trees and fish pictures. A coupled climate model program features separate model components that interact, usually through a separate, master, control program called a coupler. Each component is often developed as an individual model (like the atmosphere). The coupler or control program handles the exchanges between the different

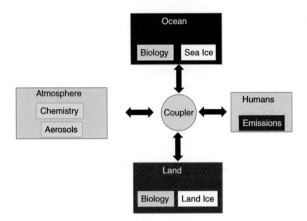

Fig. 4.9 Coupled climate model. Schematic of the component models and subcomponents of a climate model program. The coupler code ties together different spheres (ocean, atmosphere, land, biosphere, and anthroposphere) that then contain smaller component submodels (like aerosols, chemistry, or sea ice)

models. In idealized form, different versions or different types of component models can be swapped in and out of the system. These can include data models where, for example, instead of an active ocean, just specified ocean surface temperatures are used to test an atmospheric model.

The different components of the system, the different boxes in Fig. 4.9, can each be thought of as a separate computer program. There are often subprograms for different sets of processes, such as atmospheric chemistry or ocean biology. Each of these boxes can often be constructed as a series of different processes (individual boxes). The deeper one goes, the more the individual models are a series of processes.

This software construction is modular. A process is represented mathematically by a program or subroutine. It is coupled to other similar processes, like a model for clouds, or a model for breakup of sea ice. These similar models at each step are constrained for mass and energy conservation. The cloud model may consist of different processes, down to the level for a single equation, such as the condensation or the freezing of water. The processes and sub-models may be tested in some of the simple frameworks discussed earlier, and then often they are coupled with other physical processes into a component model. As shown in Fig. 4.9, the atmosphere model typically contains a set of processes for clouds, radiative transfer, chemistry, and the dynamical core that couples the motion together, as in Fig. 4.6. This is a sequence of computer codes: a set of equations, tied together by physical laws of conservation and motion.

How complicated does this get? Current climate models have about a million lines of computer code. This is similar to a "simple" operating system like Linux, but far less than a more "complex" operating system (50 million lines of code for

Windows Vista) or a modern web browser (6 million lines for Google Chrome).[7] So models are complex, but still quite compact, compared to other large-scale software projects. Of course, this level of complexity and complication means that there are always software issues in the code, that is, potential "bugs."[8] How do we have any faith in a million lines of code? Scientifically, the conservation of mass and energy is enforced at many stages: If the model is well constrained, then even a bug in a process must conserve energy and mass. Let's say that a process evaporating water is "wrong." It still cannot evaporate more liquid than is present, limiting the impact of the mistake.

From a software perspective, climate model code must be tested the same way as any large-scale piece of software. There are professional researchers whose sole job is to help manage the software aspects of a large climate model.

Climate models are constructed by teams of scientists. The teams have specialties in different parts of climate system science: oceans, atmosphere, or land surface. There are social dimensions to model construction and evolution. Some models share common elements in various degrees. This is important when constructing an ensemble of models, as one has to be careful of picking models that are very similar and treating them as independent. Models with similar pieces (e.g., the same parameterizations) may share similar structural uncertainty. Most modeling centers have a specific "mission" related to their origin and history. They focus on excellence in particular aspects of the system, or on simulating particular phenomena. It should be no surprise that model groups in India worry very much about the South Asian Summer Monsoon, or that a model from Norway has a very sophisticated snow model. This is natural. Model codes are generally quite complex. Some climate models are designed and run only on particular computer systems. Some climate models are used by a wide community. Climate model development teams typically work with friendly competition and sharing between them. Climate model developers are continually trying to improve models and always looking over their shoulder at other models. There is a negative aspect to this community, and that is "social convergence": Sometimes things are done because others are doing them. There is a desire not to be too much of an outlier.

---

[7]See the infographic http://www.informationisbeautiful.net/visualizations/million-lines-of-code/ contained in McCandless, D. (2014). *Knowledge Is Beautiful*. New York: Harper Design.

[8]The first "bug" was thought to be a result of a moth being smashed in an electromechanical relay in the Harvard Mark II computer in 1947, according to Walter Isaacson in Chap. 3 of *The Innovators: How a Group of Hackers, Geniuses, and Geeks Created the Digital Revolution*. New York: Simon & Schuster, 2014. This of course would be considered a hardware, not a software, bug but the name stuck.

4.5.2 Running a Model

So how is a model run to produce "answers" (output)? Complex climate models are designed to be able to run in different ways. Climate models can be run in a simplified way, as a single column in the atmosphere, for example. This can be done on a personal computer. But the real complexity is running every single column of atmosphere on the planet. For this, large computers with many processers are used. These are supercomputers, and they typically have many processors. How many? The number increases all the time. As of 2015, the largest machines had a million processors, or computing cores.[9] Usually a model will run on part of a machine. Since a model with resolution of one degree of latitude ($\sim$62 miles or $\sim$100 km) would have about $180 \times 360 = 64{,}800$ columns, models have several atmospheric columns calculated on one core. Note that for 15 mile (25 km) resolution, this number goes up to approximately one million columns. The atmosphere would have a certain number of cores, the ocean a given number, and so forth for all the component models. The speed of each calculation is not as important as how many cores can be used to process computations "in parallel." The total cost of a model is the time multiplied by the number of cores. More cores mean that a larger number of computations can be done in the same amount of time: The total run time gets shorter. The cost of running a climate model depends on the number of columns, and this depends on the resolution. As computers get faster and especially bigger (see below), higher resolution simulations, or more simulations, or longer simulations become possible.

The need to communicate between columns makes climate models suitable for only a special class of computer hardware. A climate model column calculated on a computer core needs to talk to the next column when the calculation of the time step is done. Thus, the system must be designed to rapidly collect and share information. Most commercial "cloud" computing systems are not designed like this. A Google search, for example, requires a computer node to query a database and then provides an answer to a single user, without communication to other cores. So only certain types of computer systems (usually supercomputers designed for research) are capable of running complex climate models efficiently.

The supercomputers used to run climate models are common now with the rise of computational science in many disciplines. Many universities maintain large machines for general use. Weather forecast centers also typically have their own dedicated machines for weather forecasts that climate models are run on. And they run on some of the largest machines hosted by government laboratories. In the United States, these machines are often found at national laboratories run by the Department of Energy. Their primary use is to enable finite element simulations of nuclear weapons: similar to the first electronic computers. These machines use on the order of 1,000–5,000 kW.[10] A watt is a rate of energy use: a Joule per second.

---

[9]An updated list is maintained as the "Top 500" list: http://www.top500.org.

[10]Data from the Top 500 list, November 2014.

A bright incandescent light is 60–100 W. Note that 1 kW = 1000 W (so we're talking about 1–5 million watts). For comparison, a large household with many appliances might barely approach 1 kW at peak consumption. Thus, the power requirement of the largest machines is the scale of a town of maybe 2000–10,000 people (assuming about two people per house). The machines themselves live in special buildings, with separate heating and cooling (mostly cooling) facilities.

For very large machines and high-resolution simulations, data storage of the output also becomes a problem, requiring large amounts of space to store basic information. Currently processing power is often cheaper than storage: It is easier to run a model than it is to store all the output. This means sometimes models are run with limited output. If more output is needed, they are run again.

So who runs climate models? Usually the group of scientists who develop a model also run the model to generate results. The groups of scientists who develop coupled climate models have grown with the different components. These groups are usually part of larger research institutes, universities, or offshoots of weather forecast agencies. The work is mostly publicly funded. These research groups are usually called modeling centers. Often, standard simulations are performed (see Chap. 11). The output data are then made publicly available. So use of model output is not restricted to those who can run the models.

4.6   Summary

Based on the fundamental principles that work every day in the world around us, a climate model seeks to represent each part of the system (e.g., atmosphere, ocean, ice, land) and each critical process in these parts of the system. Some examples include the conditions when water vapor condenses to form clouds, how much sunlight is absorbed by a given patch of land, or how water and carbon dioxide flow in and out of leaves. Each process can be measured. Each process can be constrained by fundamental physical laws. We describe many of these processes in the detailed discussion of models in Sect. 4.2 (Chaps. 5–7). The hope is that after each process is properly described and constrained, the emergent complexity of the earth system is in some way represented. With more computational power, more processes can be included. Finer grids (more boxes) can be simulated. But the broad answers should not change. As we will see in Sect. 4.3 of this book, the "hypothesis" of climate models' validity is being tested repeatedly and in many different ways.

The uncertainty that remains is considerable and is discussed in Sect. 4.3. We have discussed fundamental constraints (fundamental transformations and physical laws) on the climate system. But these constraints do have uncertainty in the complex climate system. The emergent complexity means that there are many possible states of the climate system. Just like weather can have many states in a distribution, climate is simply the average (the distribution) of those states realized in a particular finite time. The different states may evolve in response to different forcing and to uncertainties in how processes are represented in the system. We return to these uncertainties later, but now it is time for a slightly more detailed discussion of how we simulate the different major components of the climate system. Time to follow the White Rabbit a bit farther down the rabbit hole before it gets too late.[11]

Key Points

- Climate models are based on known physical laws.
- Basic processes describe the source and loss terms in equations, subject to basic laws of conservation.
- Uncertainty lies in how processes are represented (parameterized) and coupled.
- Simple to complex models exist.
- Climate models have and continue to push the limits of computers.

---

[11]See Carroll, L. (1865). *Alice's Adventures in Wonderland*. New York: Macmillan.

Part II
Model Mechanics

Chapter 5
Simulating the Atmosphere

The atmosphere is critical for understanding energy flows in the climate system. The main energy input for the climate system is the sun, and the atmosphere has an important role in how solar energy enters the earth system, and how energy leaves the earth system. As we have discussed, solar energy in visible wavelengths (shortwave radiation) mostly passes through the atmosphere and is absorbed by dark surfaces, but reflected by light surfaces, including clouds. Thus, clouds are critical for the net energy input. Energy radiated from the earth in the infrared (longwave radiation) passes through the atmosphere on its way out to space. It can be absorbed not just by clouds, but also by greenhouse gases, such as water vapor (H_2O) and carbon dioxide (CO_2). Thus, the composition of the atmosphere is critical for understanding climate.

The atmosphere is also important for connecting the different parts of the earth system. Two of these connections are related to greenhouse gases: the water (or hydrologic) cycle, and the carbon cycle. Important parts of these cycles happen in the atmosphere. The carbon cycle is critical for understanding not just CO_2 in the atmosphere, but also how carbon moves through the land surface (to be discussed in Chap. 7). The water cycle is critical for human societies and ecosystems as well as the climate system: We experience the water cycle at the surface through clouds and especially through precipitation where water hits the land surface. But water also moves energy through the earth system, and this is an important part of simulating the atmosphere. For these reasons, we start a detailed discussion of each component of the climate system with the atmosphere.

This chapter explores in more detail how the atmosphere is modeled in the climate system. We start with the different pieces of an atmosphere model: the energy flows, the circulation, and the transformation of water. This chapter contains a description of the types of atmosphere models. We also discuss the similarities and differences between models used to simulate climate and those used to simulate weather. Finally, we go into detail about some of the challenges involved in simulating the future evolution of the atmosphere.

© The Author(s) 2016
A. Gettelman and R.B. Rood, *Demystifying Climate Models*,
Earth Systems Data and Models 2, DOI 10.1007/978-3-662-48959-8_5

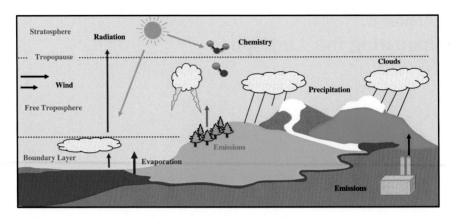

Fig. 5.1 The Atmosphere in the Climate System. Emissions and evaporation from the surface (as well as radiation) force the atmosphere from below. The sun forces the atmosphere from above. Chemistry, clouds, and wind occur in the atmosphere, along with the flows of radiation

5.1 Role of the Atmosphere in Climate

The atmosphere is a familiar part of our daily life, and the role of the atmosphere in climate features many aspects of weather that we observe every day.[1] Figure 5.1 illustrates a schematic of the atmosphere, highlighting some of the key aspects of the atmosphere in the climate system. This figure is related to the hydrologic cycle (Fig. 2.3) and the energy budget (Fig. 5.2, reprinted from Fig. 2.2). Most notably, the atmosphere is where the energy input from the sun is distributed in the climate system. It features clouds and precipitation, and with evaporation from the land surface, this represents the atmospheric hydrologic cycle. The entire atmosphere is in motion with winds that are part of a large-scale atmospheric circulation. Vertical motion is driven by buoyancy: the difference in density of air. Warm air is less dense than cold air, so it rises. As air rises, it expands and cools. But if it cools enough, water vapor condenses and forms a cloud. This releases heat, and then the air may continue to rise: This process gives rise to clouds, and to deep vertical motions in the atmosphere.

There are many variations in the atmosphere that occur in both space and time. Clouds may be only a few hundred meters in size. Temperature and winds vary from place to place, and over the course of a day. We experience these variations at quite small scales compared to the global scale, or even compared to the typical scale (62 miles, 100 km) of global models. Many clouds are small scale, and precipitation events may be very localized. These small-scale variations make it difficult to simulate and predict the future state of the atmosphere.

---

[1]A good general introduction to atmosphere and climate is Randall, D. (2012). *Atmosphere, Clouds and Climate*. Princeton, NJ: Princeton University Press.

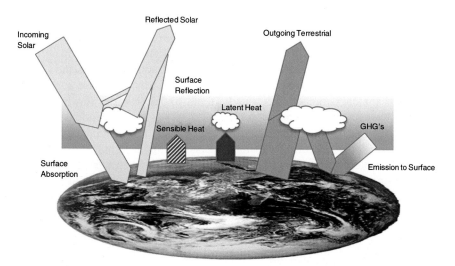

Fig. 5.2 Energy Budget. Solar energy, or shortwave radiation (*yellow*) comes in from the sun. Energy is then reflected by the surface or clouds, or it is absorbed by the atmosphere or surface (mostly). The surface exchange includes sensible heat (*red striped*) and latent heat (associated with water evaporation and condensation, *blue*). Terrestrial (infrared, longwave) radiation (*purple*), emitted from the earth's surface, is absorbed by the atmosphere and clouds. Some escapes to space (outgoing terrestrial) and some is reemitted (reflected) back to the surface by clouds and greenhouse gases (GHGs)

The atmosphere is structured into a **boundary layer** near the earth's surface, often unstable near the ground (which can warm up rapidly due to solar insolation), then stable at some level a few hundred to a few thousand feet (1–3 km) above the ground. Above this is the "free **troposphere**" (*tropos* = changing). At about 40,000–60,000 feet (12 km, the altitude at which a plane flies), is the top of the troposphere, the **tropopause**. Above this the air stops getting colder with height, and begins to warm with height due to absorption of sunlight by ozone. This region, the **stratosphere** (*stratus* = layered), is highly stable and also dry: devoid of clouds. The "weather" we experience, and most of the important climate processes, occur in the troposphere.

We have already discussed the importance of energy flows in the climate system (see Chap. 2). Many of these flows occur in the atmosphere (see Fig. 5.2). Energy comes in from the sun and is absorbed, reflected, and transmitted by the atmosphere. We think about energy flow in the atmosphere in the vertical: Sunlight hits the surface or clouds, some is reflected depending on the whiteness (albedo) of the surface, and then thermal (infrared) energy is radiated back. The greenhouse effect of CO_2 and other greenhouse gases (water vapor, methane) traps more of the thermal energy and prevents it from escaping into space, adding energy to the "Emission to the Surface" arrow seen in Fig. 5.2.

But energy also moves horizontally in the atmosphere. We see the swirling cloud patterns on weather maps and satellite images, and these large-scale weather systems move different amounts of heat and moisture around the atmosphere. Now think about what this looks like from the ground. At any one spot, nearly the same amount of energy hits the top of the atmosphere, above the clouds, where the sun always shines. The amount is broadly the same from day to day (it varies slowly with the seasons). However, the temperature and local weather (e.g., precipitation, wind) vary a lot from day to day due to atmospheric motions. This horizontal motion of energy explains why global General Circulation Models (GCMs) are good at representing climate (and weather): If we know the state of the atmosphere on a given day, we can use the basic equations of physics to estimate what the state will look like the next day, and the next. Compounding of small inconsistencies over time makes the problem of exact prediction difficult (see Sect. 5.6), but knowing the laws of physical motion and being able to conserve energy and mass helps.

Figure 5.3 illustrates why energy moves horizontally. At any one place, even over a whole band of latitude, the energy is not in balance. Energy comes into the earth system from the sun, in the form of **shortwave or solar energy**. This energy

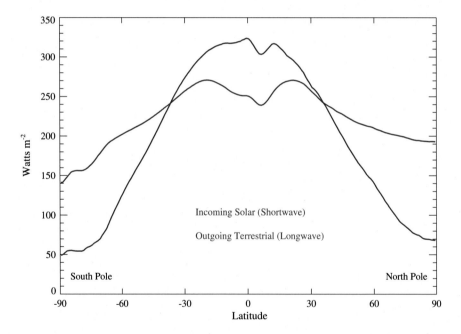

Fig. 5.3 Energy Transport. Zonal average (around a latitude circle) of the top of atmosphere energy from the sun (shortwave, incoming: *blue*) and from the earth (longwave, outgoing: *red*). The difference between the incoming and outgoing shows a surplus of energy in the tropics and a deficit of energy in the middle latitudes and polar regions. Data are from the Clouds and the Earth's Radiant Energy System (CERES) satellite. The reference is Loeb, N. G., Wielicki, B. A., Doelling, D. R., Smith, G. L., Keyes, D. F., Kato, S., et al. (2009). "Towards Optimal Closure of the Earth's Top-of-Atmosphere Radiation Budget." *Journal of Climate, 22*: 748–766

is usually at visible and ultraviolet wavelengths. The short wave solar energy input, minus any reflection, is called the net input. The net solar energy input peaks in the tropics. In mid-latitude regions the sun is lower in the sky for part of the year, and in polar regions there is no short wave energy (no sun) during polar night. The earth also radiates energy away in long wavelengths (**longwave terrestrial energy**). Longwave energy is mostly in the infrared. This is more constant and related to the temperature of the surface. As a result, there is excess energy in the tropics (i.e., at 0° latitude), and a deficit at high latitudes (e.g., at the poles), which is clearly seen in the annual average in Fig. 5.3. This gradient keeps the poles cooler than the tropics. It also means there is an energy flow toward the poles. Some of this energy is carried by water.

As we discussed in Chap. 2, water has a significant effect on the energy budget and is important for this heat transport. It takes energy to evaporate water into vapor. This energy is released when the water condenses into clouds. In the tropics, there is lots of water in the atmosphere over the oceans, and lots of sunlight. Much of this water condenses locally and drives cloud formation and deep towers of thunderstorms. But some of the water is transported long distances with the wind. It may condense far from its source: over a continent, for example, or closer to the poles. When it does so, it releases heat, as well as releasing water. This heat changes the atmospheric temperature and is also critical for driving storm systems (as with thunderstorms in the vertical dimension). As indicated in Fig. 5.2, this heat represents nearly one quarter of the total energy that hits the surface, for example, 90 of 340 Watts (W) of energy for every square meter. A watt is a standard unit to measure the rate of energy production (joule is the energy, and 1 W = 1 J/s), whether it is measuring light bulb output/usage or energy hitting the land surface. So 90 W/m^2 is the energy of a bright incandescent light bulb over 1 m^2 (about a square yard, or 10 square feet).

Water transformations are one of the more magical and complex parts of the atmosphere. These transformations drive weather and are critical for climate. Much of the poleward transport of heat in Fig. 5.3 occurs through the evaporation of water from the tropical oceans, and the movement of that water poleward, where it condenses at higher latitudes. This also works on more regional scales. The paths and events in which air tends to flow poleward from the tropics with lots of water vapor even have a name: atmospheric rivers. Figure 5.4 illustrates a picture of the earth with infrared wavelengths that correspond to water vapor. Dark areas have little water vapor, white areas have a lot of water vapor. The water vapor streams out of the tropics, feeding storms at mid-latitudes (e.g., over S. America). In the mid-latitudes, the winds blow from west to east, and carry water from oceans onto continents. One key aspect is the release of heat: As condensation begins, it heats air, which then typically rises because warmer air is less dense. The rising air then cools, is replaced by air from below (also rising), and condenses more water vapor. The water in clouds will eventually fall as rain, but the condensation adds extra heat that drives the storms and forms a critical part of the general circulation of the atmosphere.

August 10, 2015:
Water Vapor Image

Fig. 5.4 Water Vapor Image. Image of the earth on August 10, 2015 from a satellite in the water vapor band. Image from NOAA/University of Wisconsin. Dry regions are *dark*, moist regions are *gray* to *white*. Clouds are *white*. The tropics are mostly white (moist), the sub-tropics are mostly dark (dry). The high latitudes are mostly gray (moist)

5.2   Types of Atmospheric Models

In Chap. 4, we discussed the level of complexity of models. There exists a hierarchy of models used to simulate the atmosphere, illustrated in Fig. 5.5. A **box model** is a simple model that has a single or small number of boxes (Fig. 5.4a). A box model has inputs and outputs and one temperature. It is used to model the energy balance of the climate system. A simple energy balance box model of the earth, with one uniform temperature, essentially assumes a uniform atmosphere.

More common are **energy balance models** with a several-layer atmosphere in a column (Fig. 5.5b). These generally have a realistic variation of temperature with height when the flows of energy from the top of the atmosphere to the surface are taken into account. Energy balance models are useful for examining changes in the composition of the atmosphere. Typically they exist solely in a single atmospheric column, designed to represent the whole planet or a region of the planet.

Energy balance models are a type of **single-column model** (Fig. 5.5b). The representation does not include geographic variation. Sometimes (as with energy balance models), the intent is to understand energy flows. But single-column models can represent any number of complex processes, except that they do not have horizontal motions. In Fig. 5.5b there are movements of mass and energy through a top and bottom boundary (as from the sun at the top and the land surface below), and there is a forcing at the side boundary (often an imposed wind speed),

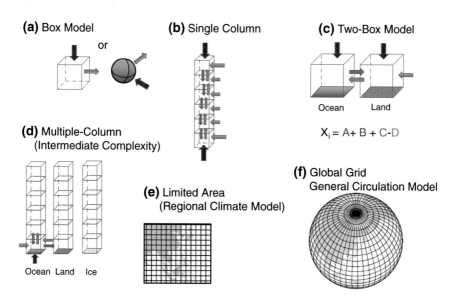

Fig. 5.5 Hierarchy of Models. Different types of atmosphere models. **a** Box model (zero dimensions). **b** Single column model (one dimension in the vertical). **c** Two box model (zero dimensions). **d** Multiple column model. Sometimes multiple column models are intermediate complexity models with columns representing a region like a country, so 2 dimensional: one dimension in space, one dimension in height. **e** Limited area or regional climate model (three dimensions). **f** General circulation model (GCM) on a global grid

in this case varying with height. Such a model might be used to estimate the surface temperature given solar radiation and different greenhouse gases, or to simulate cloud and precipitation processes.

Often several different boxes or column models are tied together, and one box affects another. For example, a simple two-box model might try to represent the surface of the earth with a box representing the atmosphere over land and a box representing the atmosphere over ocean, as shown in Fig. 5.5c. Or there might be one box for the tropical regions and one box for the polar regions. These models may contain some sophisticated processes to represent the flow of energy, mass, and cloud formation that describes regional temperature and precipitation.

Sometimes a small number of boxes or columns are used specifically to represent conditions over large regions of the planet. There might be one column for each continent, and one column for each ocean basin, and one column over ice. An example is illustrated in Fig. 5.5d. The boxes are in balance internally and with each other, and they can exchange with each other and the boundaries. Such models are often called **intermediate complexity models**.[2] These models might have 20 or so

[2]Claussen, M., Mysak, L., Weaver, A., Crucifix, M., Fichefet, T., Loutre, M.-F., et al. (2002). "Earth System Models of Intermediate Complexity: Closing the Gap in the Spectrum of Climate System Models." *Climate Dynamics*, *18*(7): 579–586.

columns over major land masses, to represent temperature and precipitation, that then drive economic models in different countries or regions. The oceans might be represented by one column for each ocean. Intermediate complexity models capture some of the basic energy budget components and limited circulation. These models are used when the needed climate variables are simple. Simple outputs might be the average surface temperature or precipitation in a region or over an entire country. These outputs can then be used to efficiently drive a country-level economic model.

Finally, we come to models that have finite elements covering the earth. These can be regional (a regional climate model[3]) or global. **Regional climate models** or **limited-area models** have boundaries: They do not represent the entire surface of the earth (see Fig. 5.5e). The limited area enables them to be run with finer resolution than a global model. Finer horizontal resolution is good for representing the effects of surface features like mountains (topography). If you want to understand the climate of a region near or within mountain ranges, it is critical to represent the effects of the mountains correctly. This may be easy to understand in terms of a small region like a mountain valley, or the region on the dry side of a mountain range away from the coast, such as eastern Oregon in the United States, or the high deserts on the Andes mountain range in Peru and Chile.

Limited-area models may also include more processes because they represent smaller regions with fewer boxes, a limited **domain**. The difficulty is that they have "edges" or boundaries: Air must blow into and out of them, along with the energy associated with the air, and water in vapor or in clouds (the "blowing" around is also called **transport**, or advection). Limited-area model boundaries (the region just outside the domain of the model) have to be defined from somewhere to determine what values are given to the model at the boundary. For examining present-day climate, the conditions at the boundary can be taken from observations, often using data collected for large-scale weather prediction. Indeed, many fine scale models of weather have limited domains. For climate prediction, limited-area models are difficult because the boundaries mean energy and mass can leave the model (nonconservation).

It is a challenge, however, to use these models for the future, since they need to have specified future boundary conditions. The results are often or usually strongly dependent on the boundary conditions specified. But these models, with their small grid spacing and detailed representation of processes, are good for representing details of local climate variation through weather events: extremes of precipitation and temperature, for example.

Models with a global grid are good for representing the overall patterns of motion of the atmosphere (see Fig. 5.5f). They have no horizontal boundaries. These global grids are known as **General Circulation Models** (**GCMs**). They need only a top (space) and a bottom (the surface of the earth) boundary condition. The

---

[3]Rummukainen, M. (2010). "State of the Art With Regional Climate Models." *Wiley Interdisciplinary Reviews: Climate Change*, *1*(1): 82–96.

bottom boundary is fixed, and the top boundary is generally fixed at some level between 25 and 50 miles (40–80 km) above the surface. This is well above the top of the troposphere. At these boundaries it is much easier to enforce the basic principles of conservation of energy and mass that are fundamental constraints on climate. If the top boundary is defined as a constant pressure, and air cannot flow through the bottom boundary, then the mass of air must be conserved. Energy can flow in and out at the top and bottom, but otherwise must circulate and move from one box to another. Because of these conservation properties, it has been natural to turn to GCMs for understanding and predicting climate. Conceptually they allow conservation to be achieved. In practice, this needs to be done carefully, but generally it is a tractable problem. Conservation is important because changes to climate result in small changes to the energy budget, and in a conservative model, the change in the energy budget would make the heat go somewhere: like into the thermal energy (temperature) of the surface.

5.3   General Circulation

So what does this general circulation look like? By **circulation**, we mean air motions that eventually must return to their starting point (circulate) because there are no horizontal boundaries, and there are vertical boundaries to the atmosphere. Because the air is not created or destroyed, it has to go somewhere, and other air takes its place; hence, it circulates. The earth has general circulation patterns (hence the term General *Circulation* Models). The general circulation gives rise to the basic distribution of rainfall and temperature: wet, dry, hot and cold regions, and, as a result, vegetation patterns (see Fig. 5.6). The easiest way to see the general circulation is to simply look at the earth from space. The tropical land regions within about 15° latitude of the equator are mostly dark and green: West Africa and the Congo, Indonesia and the Amazon basin are tropical rainforests, or what we call jungles. In the tropics, sunlight drives evaporation, and winds come together to cause air to rise and the water to rain out. Then the air sinks on either side of the equator in the subtropics. Rising air gets colder and promotes condensation and cloud formation; sinking air is dry, and does not form clouds or rain. The dry regions are dark in the water vapor image of Fig. 5.4, which was taken at the same time as the visible image in Fig. 5.6. In Fig. 5.6 you can see the brown, dry desert regions at the same latitudes on either side of the equator. The desert regions of North America are the same latitude ($\sim 30°N$) as the Sahara in Africa, and in the Southern Hemisphere, the deserts of Australia are at the same latitude south of the equator ($\sim 30°S$). This is also the latitude of the Atacama Desert in South America (see Fig. 5.6). At higher latitudes, the land gets green again in the **middle latitudes** (40°–60° north and south of the equator): Europe, North America, and Asia. Then, of course, in polar regions a new color emerges: white for snow, ice caps like Antarctica, and sea ice–covered ocean.

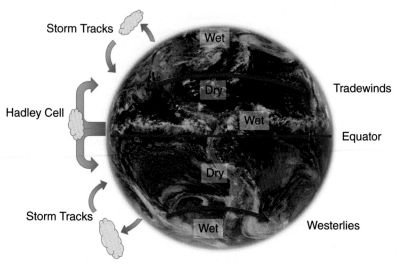

From: NASA/NOAA   Aug 10, 2015 Visible Image

Fig. 5.6 General Circulation of the Atmosphere. Visible image of the earth with the general circulation overlaid. The image is for the same time (August 10, 2015) as the water vapor image in Fig. 5.4. Wet regions with rising motion around the equator in the upward branch of the Hadley Cell. Tradewinds blow westward in the tropics (easterlies, from the east). Downward motion in the dry regions where the deserts lie on either side of the equator in the subtropics. Then it is wet again in the mid-latitude regions of the storm tracks with eastward winds (westerlies, from the west)

These vegetation bands generally result from upward and downward motion, conditioned by the basic laws of physics that govern the climate system (see Chap. 4). The exact latitude is a result of the energy input, and the rotation speed of the earth. Air moves upward in the deep tropics, causing clouds to form. It circulates poleward at upper levels into the subtropics. There air cools and descends in the desert regions of the subtropics. This is known as the **Hadley circulation**, after George Hadley, who came up with the basic framework of why the "trade winds" blow westward near the equator.[4] Another broad cell extends over mid-latitudes, with rising motion there and sinking motion over the poles. The same effect creates the banded cloud patterns seen on other planets such as Jupiter and Saturn: bands of different colors, representing clouds at different altitudes due to the motion of the atmosphere in cells up and down, as on earth. GCMs exist for other planets as well, and these same equations can produce a tolerable representation of the banded cloud and wind structure on Jupiter, another useful test of the basic physics in a climate model.

Although the latitude bands are a guide, the circulation is not strictly by latitude: Not all regions at the same latitude have the same climate. Denver, Madrid, and

---

[4]Hadley, G. (1735). "Concerning the Cause of the General Trade-Winds" *Philosophical Transactions, 39*(436–444): 58–62.

Beijing all lie at the same latitude (40°N), but they have very different climates. Dry and moist regions are localized regionally, due to the contrasts between the land and ocean. Denver and Madrid are at relatively high altitudes, and Denver is in the middle of a continent, whereas Beijing is on the eastern edge of a continent. These geographic differences give rise to different climates.

In the tropics, upward motion of air is favored where it is warmer over land, and where the ocean basins are warmer. Air moves up in some regions, and down in others. Warm water near Indonesia and tropical land masses over Africa and South America have preferentially upward motion, with downward motion favored in the eastern Pacific. The tropical circulation pattern of upward and downward motion in particular regions is called the Walker Circulation after Sir Gilbert Walker, the longtime head of the Indian Meteorological Department in the early 20th century, who first charted many of the tropical circulations. Air also moves poleward at upper levels due to this heat input in the tropics, and eventually it descends in the subtropics.

Most modern GCMs do a good job of broadly reproducing these different dry and wet regions, or **climate regimes**. The driving forces are from the largest scales: land and ocean contrasts, the rotation of the earth. But getting the details right is critical for understanding how things evolve and will change at any given place. If the locations of the regions of precipitation (such as the edge of the tropical wet region) is off by a "small" amount (a few hundred miles or kilometers), this may mean a vastly different climate at a particular place. Billions of people live in the subtropics, which lie both north and south of the equator (between $\sim 10°$ and $\sim 30°$ latitude). The subtropics include large swaths of India, Southeast Asia, China, Africa, and North America. We discuss the potential shifts in climate regimes further in Chap. 10, when we consider uncertainty. But society's vulnerability to climate—or to say it positively, the extreme adaptation of human populations to climate—makes understanding and being able to simulate climate critical.

5.4 Parts of an Atmosphere Model

So how is a GCM constructed? The basic concept involves coupling the different physical processes (clouds, energy flows, exchanges with the surface) with the basic laws of motion as discussed in Chap. 4. Some of these processes are transformations (like clouds), and some are external processes that push (force) the model. The laws of motion provide the resulting distribution of winds (motion, kinetic energy) and temperatures (thermal energy). The motion moves water vapor, clouds, and chemicals in the air. Physical and chemical processes describe transformations that determine the sources and loss processes of water vapor, cloud water/ice, and chemicals. Loss processes are sometimes called **sinks** for the water that flows down a drain. As described in Chap. 4, the first step is to calculate all the processes and determine the sources and sinks of critical parts of the atmosphere (water, clouds, chemicals). Each process in each box is parameterized, shown in Fig. 5.7a. The result is sources and sinks, along with energy transfer by radiation and by surface fluxes (shown in Fig. 5.7b). These sources and sinks are then used to "solve" the equations of motion, energy, and mass conservation on a rotating sphere (Fig. 5.7c). This step provides winds and temperatures that can be used to recalculate the processes; hence, the model marches forward in time.

As discussed in Chap. 4, an atmosphere model is a component of the climate system. It contains a representation of the equations of motion and conservation (Fig. 5.7c), as well as a representation of physical processes in the atmosphere (Fig. 5.6a, b). These physical processes must be represented in individual boxes (Figs. 5.7 and 4.3) at each location on the planet based on a grid of points (Fig. 5.7).

Now think for a second about the global grid in Fig. 5.7. The latitude-longitude grid has about 30 different latitudes (or a resolution of about 6°). At this scale, there are four grid boxes that encompass all of Japan, with each including part of the ocean around it. A typical modern model has a finer grid than this: maybe 1 or 2 degrees of latitude. But that is still 68 miles (110 km) on a side. Now think about a sky: The entire sky you can see from near the ground (from a hilltop or a tall building) in all directions is not much more than that distance. So if that represents one column of air, with one value for each layer in it, what happens when the sky is partly cloudy? What happens when there is a thunderstorm that occupies only part of the grid box? What happens when it rains in only part of a grid box?

More generally, what do we do with variations across the grid box, when one number will not do? Many processes in climate models have ways of dealing with this below-grid-scale ("sub-grid") variability. It is one of the most difficult problems of all scales of modeling: What do you do with variations that occur below or even near the grid scale?

The sub-grid scale problem is mostly variations with scales *near* the grid scale. If the grid is 62 miles or 100 km (1° latitude), then sub-grid variations are greater than 0.6 miles (1 km). Very fine scale variations (much less than 1 km in this example) can often be represented with statistics, because the scales of interest are well separated from the grid. A 100-km grid has 10,000 one kilometer square elements (10,000 km^2). It has 1 million elements that are 100 m by 100 m (about two American or European football fields next to each other). Often we can represent a distribution of elements (a probability distribution function of small-scale features in space). But cloud systems like thunderstorms are often 5–20 km on a side and there might be just a few small-scale features in a grid box. So it becomes difficult to generate good statistics with only 25 elements in a 100-km grid box. Even at the 1 km scale, clouds will vary quite a bit. Even over the football field scale (300 ft, 100 m), small clouds have important variations. But they may be captured statistically since they are now quite smaller than the grid scale, and there are 1 million of them in 10,000 km^2.

If the statistics are not well known, because there are a few elements due to large size or rarity, then the statistics are not well sampled. This often happens when the scales of variability are not well separated from the grid scale (they are a significant fraction of the grid size). This is particularly a problem for clouds in the atmosphere, and clouds are particularly important for climate.

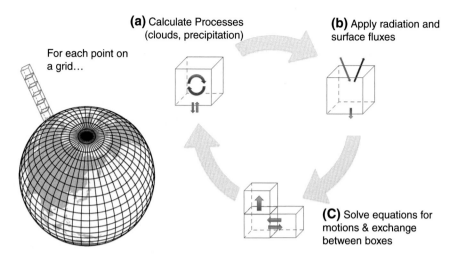

(a) Calculate Processes (clouds, precipitation)

(b) Apply radiation and surface fluxes

For each point on a grid...

(C) Solve equations for motions & exchange between boxes

Fig. 5.7 General Circulation Model. Schematic of calculations in a time step in each grid box of a General Circulation model, including. **a** Calculate processes, **b** Apply radiation and surface fluxes and **c** Calculate motions and exchanges between boxes

We return to this problem of **subgrid variability** later, but for now it is apparent that this variability often makes determining the tendency and forcing terms in models very complex, and it is one of the central challenges to climate modeling. It is an acute challenge for all components, but especially for the atmosphere. We often can only approximate the terms (statistically, for example) and cannot represent them explicitly: Hence, we build parameterizations or submodels of the different processes (also based on physical laws) and link them together.

Figure 5.7c shows the step where the changes due to all the different processes are applied to basic physical equations. These equations are used to determine the winds and temperatures at every model grid box that will result from all the different processes, like radiation and clouds. This is often called the **dynamical core** of the model (Fig. 5.7c). The laws of physics and fluid motion describe what happens to a compressible fluid (air) when it is pushed, or heated. The temperature of the air helps determine where there is higher pressure (more density, often colder) and where there is less density (warmer). The equations describe how air tends to move from higher density to lower density. The wind we feel is just air motion.

Changes in water (condensation, evaporation) are important in altering the heat content of the air (temperature). Other forces that pull on the air include **drag**, or friction from the surface (more over mountains and rough terrain than over smooth terrain, and varying with vegetation). This list of forcing terms is put into a set of equations to determine the response of the air to these changes (e.g., in winds or in temperature). Conservation of mass and energy are also applied. The winds are used to transport substances in the air: cloud drops, water vapor, and other substances such as ozone or small particles like dust (small particles in the air are known as **aerosol particles**).

Each of these substances then has a balance equation. Think of a box with a pile of balls in it, with each ball representing a molecule (or many molecules) of water, dust, or a chemical. Some are pushed in and out of the box in different ways (moving to a box adjacent in the horizontal, or moving up or down in the vertical), and some balls may move out of the bottom of the atmosphere (or come from the surface into the atmosphere). Some times the balls change category: from water vapor to cloud water for example. But all the balls need to be accounted for using the calculated wind and the laws governing fluid flow. The laws take into account the rotation of the earth (which affects the weather patterns on large scales, by altering the momentum of the air).

Let us put this all together and walk through a single time increment, or time step, in an atmosphere model. The basic structure of an atmosphere model is to break up the atmosphere into pieces representing a part of the earth's surface. For each grid box on the earth's surface, there is a column of air (see Fig. 5.7). The processes or forcing terms due to clouds, precipitation, chemistry, and turbulence local to a grid box are estimated using parameterizations (Fig. 5.7a). The physical laws for radiative transfer (the flow of energy) and conservation of mass are applied locally to the column (Fig. 5.7b). Then the changes in the local quantities are applied as *forcing terms* to equations that describe the motion of air on a rotating sphere, or the dynamics of the model (Fig. 5.7c). The equations provide estimates of the change in temperature and wind, in addition to the changes of each substance such as water or cloud drops. The updates are applied and the process begins again.

So what are these forcing terms that push the model dynamics? They are the input (and extraction) of heat, and the changes (and transformations) in the different chemical species, such as water vapor, represented in Fig. 5.6a, b. In an atmosphere model, we need to understand the sources and sinks of water vapor and the flows of heat in the system that drive the laws of motion. These forcing terms are made up of different physical processes: each one itself a complex set of different effects. The ones we discuss in the sections that follow are clouds, radiative transfer, and chemistry.

5.4.1 Clouds

Clouds are probably the single most complex and important part of representing the forcing terms in an atmosphere model. This is for two reasons: First, clouds are white. Clouds are white due to the size of drops absorbing uniformly across wavelengths. The presence of a cloud over a usually darker surface alters the energy input into the system. Second, clouds precipitate, and precipitation is critical for plants and the land surface (in addition to moving heat around). Clouds are also incredibly complex (and beautiful). Needless to say, all those wonderful shapes we see in the sky are well below the grid scale of any global model, and the interactions of clouds and their environment are incredibly complex. The environment around a thunderstorm, or a line of thunderstorms, contains complex motions at small scales that are impossible for global models to represent. We typically use observations and smaller-scale models to reduce the complexities, such as those in clouds, to a series of parameters we can represent in a *parameterization* (see box in Chap. 4). There are several types of parameterization of clouds: usually one for layered clouds (stratus) and one for vertically deep and cumulus clouds (including the cumulonimbus clouds that are thunderstorms). There are also special representations of clouds and the turbulence near the ground in the layer at the bottom boundary of the atmosphere.

At the many-kilometer scale used in a global model, the small-scale details of a single cloud cannot be defined by a single number. These details include the speed of rising air, the distribution of cloud drops, or the distribution of raindrops. However, the effect of the cloud motions on the model can be discerned. The cloud motions transport water substance, including precipitation, and the cloud changes the radiative transfer in the atmosphere (see Sect. 5.4.2). These effects are subject to large-scale constraints (e.g., energy and mass conservation, local winds and temperatures at each level) that help us constrain the problem and build a cloud parameterization to represent the effect of the clouds in the 15 min or so of a climate model time step.

The simplest cloud parameterization comes from the observation that excess water condenses when air reaches its saturation vapor pressure of water molecules for the gas phase (water vapor). Described in the 19th century and known as the Clausius-Clapeyron equation, it can be used as a simple representation of a cloud. If the air has more than the amount of water that can be vapor (gas) for a given temperature and pressure, then the rest of the water becomes liquid: and becomes a *cloud*. Though crude, this representation is a basic check on modeling. Applying subsequently more "rules" allows more complexity (ice is more complex, for example) and better realism and representation of the appropriate complexity.

The simple Clausius-Clapeyron equation will not make a thunderstorm or a line of thunderstorms. This requires more computations, and more computer time, of course. The decision then becomes how complex a model is desired, and much of that complexity flows from the representations of processes. Some are not included at all, and some are included in different levels of detail, depending on the model's purpose. For example, if you want to predict air quality and local air pollution, then better representations or parameterizations of chemistry are required than if you want to just predict the surface temperature.

Clouds are also the process responsible for precipitation. From a climate perspective, precipitation is energy. Energy evaporates water from the surface, and then the energy is released when water condenses to form clouds. Eventually precipitation returns this energy to the surface of the earth. Each mass of water has a unit of energy needed to evaporate it, the "latent heat" of evaporation. On a global scale, the total precipitation from clouds must equal what is evaporated from the surface. This conserves water. The limitation for climate is how much energy at the surface can evaporate water. Thus, clouds and precipitation link the hydrologic cycle and energy budget of the atmosphere. And the cloud parameterization must account for all the transformations.

5.4.2 Radiative Energy

Beyond the several different types of cloud parameterization, other important processes in the atmosphere must be represented. We have already discussed the flow or transfer of energy in the climate system. Accounting for all of the energy in the system is one of the great strengths of climate models. To do this, there is a representation of how radiative energy moves around and is converted. Accounting for the transfer of radiative energy (radiative transfer) is another important parameterization (see Fig. 5.7b). Radiative transfer parameterizations typically include many of the elements seen in Fig. 5.2: the input of energy from the sun; absorption and reflection in the atmosphere; and emission from the atmosphere, particles, and the surface. The physical laws governing the transfer of radiative energy were discovered in the 19th century. Transfer of radiative energy is mostly the motion of photons (electromagnetic radiation). This is the same branch of physics that describes how electricity and wireless communications work. The science is fairly straightforward. But clouds complicate the transfer of radiative energy in the atmosphere. In a clear sky, the radiation is fairly uniform (it varies with the surface type). The variation of clouds at small scales changes the radiation a great deal: Think of how the temperature varies when the sun goes behind a large cloud.

As we have seen, greenhouse gases are an important contributor to the energy budget. The greenhouse gas that varies the most spatially (including in the vertical) is water vapor, and representing it correctly is a challenge. The physics of water vapor absorption and emission itself is actually quite complicated and must be parameterized. But the most complex part of understanding the flows of radiative energy is clouds. Water vapor mostly affects the longer wavelengths (those emitted by cooler bodies like the surface), whereas clouds affect both the longwave and the shortwave (solar) wavelengths. The goal of a radiative transfer parameterization is to take the distribution of clouds, water vapor, and particles, along with surface properties, and represent the impact of solar and terrestrial radiation. These form terms that force the equation of thermal energy in the dynamical core.

5.4.3 Chemistry

Chemicals and atmospheric chemistry affect climate in a number of ways. Some are important greenhouse gases (CO_2, methane), some are important for air quality and can damage human and plant health (low-level ozone), and some block harmful ultraviolet light from reaching the surface (high-level ozone). Note that ozone is listed twice. Ozone in the lower atmosphere near the surface is bad (highly reactive and damaging to living tissue), ozone at high altitudes is good (absorbing the ultraviolet light that causes sunburn and skin cancer). **Ozone depletion** refers to reductions of ozone at high altitude in the stratosphere (which is a bad thing). **Photochemical smog** increases ozone near the surface (also a bad thing). Smog is also made up of particles. A few of the particles are natural: organic material from plants and dust. Most smog particles are human-made, including soot and partially reacted emissions from fossil fuel burning. Technically these fossil-fuel-derived particles are unburned hydrocarbons and volatile organic compounds (VOCs).

Understanding these different chemical "species" (like an ecosystem full of animal species) is important for several aspects of climate. Chemical species affect the absorption and radiation of energy, so chemical species can alter the energy budget. Chemical species and especially aerosol particles can change cloud properties: Particles help "seed" clouds, so changing their distribution and type can change clouds and precipitation. Chemistry also affects air quality near the surface.

To simulate the chemical transformations in a climate model, each different compound is represented, and each chemical reaction between species is described and simulated. Chemical parameterizations represent from a few to a few hundred different species or compounds: like ozone (O_3). These parameterizations describe chemical reactions between species and between species and their environment. For example, $O + O_2 \rightarrow O_3$. Here O is atomic oxygen, O_2 is stable oxygen gas and O_3 is ozone. How fast the reaction occurs depends on the amount of each reactant compound (left-hand side, O and O_2 in this case) and the reaction rate. Chemical reaction rates depend on temperature, in addition to the quantities of different species that react. Some chemical reactions are driven by sunlight, such as the break up of oxygen gas: $O_2 + \text{sunlight} \rightarrow O + O$. The reaction rates are measured in laboratories. A parameterization of atmospheric chemistry may have several hundred or several thousand reactions that need to be solved at the same time. The entire process is another form of parameterization in models. Chemistry also occurs in other components of the climate system (land and ocean, especially), which we discuss in Chaps. 6 and 7.

These various terms, chiefly those for clouds, radiation, and chemistry are used to "force" or push the atmosphere. The terms are fed into the equations in the dynamical core and used to change the state (i.e., winds, temperature, condensed water, and the quantity of different species) in the atmosphere. The parameterizations depend on each other in complex ways. Here are a few examples: Clouds move chemical species. Aerosol particles are determined by chemical reactions, and these particles can alter the number of cloud drops. Chemical reactions are affected by the amount of solar radiation (the number of photons). The radiation affecting chemistry is altered by clouds. This makes the individual grid boxes in an atmosphere model more of a web than a linear circle of processes.

5.5 Weather Models Versus Climate Models

We discussed in Chap. 1 some of the differences between climate (the probability distribution) and weather (the location on the distribution). So what is the difference between a weather model and a climate model? The main difference is in the way the models are set up and run (integrated forward in time). The climate model starts up with some state of the atmosphere. For "climate" purposes, the initial state is not important: If you are looking for the distribution of possible precipitation events in 50 years, then it is independent of the initial state. But if you are looking for the distribution of precipitation events in space at a particular time (like tomorrow), then the starting point is important.

Both weather and climate models use similar equations, and similar processes. Because the simulation time is shorter, weather models can be more complex. Weather models often have finer horizontal resolution to better resolve topography and often contain more details of processes. They are often just regional models. If you are predicting the weather only a few days in advance, knowledge of the winds at the edges of the model at the present time is often sufficient. They also need not have absolute conservation of energy and mass, since small leakages over a few days do not affect the weather (if they are not related to weather events, like clouds). In this respect, many weather models have more detailed processes at higher resolution, but they are not good climate models because they do not conserve energy or mass.

A key difference between weather and climate models lies in how they are started, called the **initialization**. For short-term weather, it matters very much that the state of the atmosphere is as realistic as possible, and for this, weather forecasters worry a great deal about how observations (from surface stations, weather balloons, and satellites) are fed into the model. The evolution of weather systems depends on accurately understanding the state of the atmosphere. Errors in the description of the temperatures and winds contribute to errors in weather forecasts. So great care is taken to minimize the errors in the observed state. Errors are minimized by correcting observations and by taking into account as many observations as possible. In specific cases, new observations are taken in critical areas. A good example of additional observations are the "hurricane hunter" aircraft that fly around tropical cyclones (hurricanes in the Atlantic) and take additional observations. These observations are fed immediately into weather models to try to better predict the near-term evolution of a particular storm by reducing the uncertainty in its current state.

There is a spectrum of forecasts, ranging from short-term weather (1–5 days), to medium-range weather (7–12 days), to "seasonal" (next 3–6 months) or "interannual" (1–3 years) to "decadal" (5–10 years) forecasting. Beyond the short-term weather scale, most of the work is done by carefully initializing global climate models with observations. Basically, climate and weather models share similar sets of equations, and as computation power increases, they look more and more alike. Climate models are run at higher resolution more typical of weather models, and some weather models are run longer (7–12 days or even months) to do seasonal forecasting. The big difference is in the initial conditions: They are very important for weather models, but not so important for climate models. Climate models, if they are set up with detailed observations similar to weather models, do a comparable job of predicting the weather. And each set of models learns from the others. Climate models are adding more detail to their process representations (like cloud parameterizations), often using techniques from weather forecast models, and weather models are using techniques for conservation of mass and energy and transport schemes to better predict events longer than 5 days away. These parameterizations (such as for radiative transfer) often come from climate models.

5.6   Challenges for Atmospheric Models

The discussion of an atmosphere model here has been very reductionist. You would assume that simply getting each process correct, and adding them all together with sufficient resolution, would produce a reasonably correct result and that it is simply an exercise in "turning the crank" forward following the basic laws of physics. But this, of course, is not the case, and large uncertainties are present in simulating the atmosphere. These uncertainties come from several different sources: uncertain and unknown processes, representing scales, and the complication of feedbacks and interactions between processes in the system. Many of these challenges are also important for the ocean (Chap. 6) and the land surface (Chap. 7).

5.6.1   Uncertain and Unknown Processes

The processes that occur in the atmosphere are often very complex. So representing the different source and sink terms can be highly uncertain. The uncertainties typically might introduce long-term errors into climate simulations. For example, consider the estimation of the radiative flows in the atmospheric column. If the surface has too much snow cover, then the albedo would be incorrect. The bright surface would reflect the wrong amount of heat. This might lead to a long-term error in temperature (a temperature bias). There may also be fundamental errors in the representation of physical processes. For example, the absorption and emission by a particular atmospheric gas such as water vapor might be incorrect. This would result in energy from the sun being absorbed at the wrong level in the atmosphere. It might create a temperature bias there, or cause the wind to blow in a different direction. These fundamental errors in the representation of physical processes might be small. But some of the climate signals are small as well, and these errors may be important. The results of a series of processes that have errors sometimes work against each other: If water vapor absorbs too much, then the surface might reflect too much, and the resulting energy budget may work out correctly. These are often termed **compensating errors**.

The goal of model development is to use more detailed models of the processes and detailed observations to try to test each process separately and minimize these errors. This has been very successful, for example, in testing the representation of the radiative transfer processes. For radiative transfer in the absence of clouds, the basic physical laws provide a good theory and equations for electromagnetic radiation. We also have good "scale-separation" between the scale of the radiation process at the level of atoms and particles and the atmospheric scales of grid boxes. The theory tells us that the large number of atoms in a model grid box will behave in one way under given conditions, and so the effects of radiative energy flows at the large scale can be well described.

5.6.2   Scales

This discussion brings us back to the concept of scales as a source of uncertainty. Representing processes at different scales is incredibly challenging. Although there is good scale separation for radiation in the absence of clouds, the addition of clouds to the problem makes it more complicated. Uncertainty in the clouds themselves is even more of an issue. Clouds are variable on typical atmospheric scales, and they vary tremendously on the scale, or resolution, typical of global models (1–60 miles, or 2–100 km). The motions and interactions happen at small scales but are not uniform as they are for electromagnetic radiation. Electromagnetic radiation operates on scales of atoms, and the large number of atoms have some uniformity that allows us to describe their collective behavior with precision. Clouds are the result of a threshold effect: Add a bit more water vapor, or cool the temperature, and suddenly a cloud forms. Condensation drives changes in heat that generate complex and fascinating cloud systems. The atmosphere can go from a clear sky to a thunderstorm or even a tornado in a matter of hours, within the same air mass, with gentle forcing from the surroundings. These complex motions at small scales are hard to represent. On a global scale, tornadoes themselves may not matter much, but similarly, small scales can have significant effects on climate. For example, the altitude that thunderstorms reach is important for mixing chemicals in the atmosphere, and even heat. Tropical cyclones are a significant mover of heat between the atmosphere and ocean, and to high latitudes, and this is dependent on small-scale cloud processes.

Furthermore, it is the extreme and rare events that drive climate impacts, so representing the extremes (of precipitation amounts, or of the variability of precipitation) is critical for climate impacts. And these extremes often depend on the smallest cloud scales that have to be heavily parameterized. For example, assumptions about the sizes of raindrops can have significant implications for the intensity of rain, and even the overall presence of water suspended in clouds, with large effects on climate.

5.6.3   Feedbacks

In addition to uncertainties in individual processes, there are uncertainties in the interactions between these processes. One process affects another, and the second process feeds back on the first. In particular, when the climate system is forced, the feedback loops alter the response of the atmosphere. There are many of these feedback loops in the atmosphere. For example, thinner clouds may allow more radiation to be absorbed at the surface. This may warm the surface. Thus, fewer clouds form. This is an example of a cloud feedback. The feedbacks from clouds are in fact one of the largest uncertainties in the climate system, because they alter the total energy absorbed by the earth (remember, clouds from above are white, and they tend to reflect more radiation away from earth than the underlying land or ocean surface in their absence). Water vapor has a large positive feedback that is more certain. Warmer air holds more water vapor, and more water vapor absorbs more heat (a thicker greenhouse blanket over the earth), which warms the planet more, and so on.

This interplay of feedbacks is what makes climate prediction even more complex than weather prediction in some ways. Feedbacks that govern the overall temperature, and the distribution of heat and moisture (especially precipitation) are especially critical for climate. Feedbacks are often hard to observe, especially on larger scales, such as cloud and water vapor feedbacks. Feedbacks are emergent properties of the climate system and arise from the interactions in the system. This makes **validation**, or verification, of the processes and their interactions difficult.

A host of feedbacks in the atmosphere may affect climate response to higher greenhouse gas levels. The feedbacks respond to the changes in temperature caused by absorption of more thermal energy by the greenhouse gases. One is the water vapor feedback. Water vapor is the largest greenhouse gas, and a warmer world has more water vapor, as noted earlier. Another is a large negative feedback. When the temperature warms up, the earth radiates more energy to space, attempting to cool itself off. The hotter the temperature, the more it radiates heat away. This is a negative feedback that stabilizes climate and allows the temperatures to come into balance. The most uncertain climate feedback in the atmosphere is the response of clouds to climate change.

5.6.4   Cloud Feedback

Clouds are the biggest contributor to changes in albedo (surface reflectance). Clouds can both warm and cool: Low clouds act like a reflector (cooling the planet), and high clouds, in addition to reflecting, also act as a blanket, trapping thermal energy from escaping (warming the planet). High clouds are cold, and absorb radiation from below, transmitting it both upward to space and downward to the surface, so that more radiation remains at the surface. Low clouds however emit at the same temperature as the surface so they do not have a large longwave effect (the low cloud effect is for shortwave radiation).

The net effect is a cooling (low clouds and high clouds both cool). Unlike snow and ice, clouds are found everywhere and over large regions of the planet (about two-thirds of the planet, on average). The cooling effect of clouds is 10 times larger than the current human-caused climate forcing.[5] Thus, small changes in clouds can mean big changes in energy in the earth system. Clouds are ephemeral, variable, and very small scale, which means predicting their evolution in any scale of model is difficult, let alone predicting small changes in their averaged ability to reflect sunlight.

The cloud brightness (albedo) can be changed also without changing the extent of a cloud: With the same area, a cloud can be brighter or dimmer (think about dark thunderstorms, or thin, nearly transparent cirrus clouds). Small changes in the particles in clouds (their number or size) can also change the brightness. And different clouds may respond in different ways, so that "cloud feedback" is a sum of cloud changes. In the tropics, increased temperatures may make deeper thunderstorms and more high clouds that warm, whereas in polar regions, warmer temperatures may make more liquid clouds that are brighter and longer lived than ice clouds and they cool. The regions where clouds are most sensitive are also uncertain, but it is thought to occur in regions of extensive low clouds at the edges of the subtropical dry regions. This range of scales for cloud feedbacks, and the difficulty of observing the aggregate (total) impact of clouds over the globe, makes predicting cloud feedbacks difficult. One of the central reasons for the different climate sensitivities among climate models is the uncertainty due to cloud feedbacks. The sum of all these cloud feedbacks is thought to be positive, but the magnitude is uncertain.[6]

---

[5]Small changes to the cloud radiative effects can mean significant changes to the energy absorbed and transmitted: clouds cool by about 50 W for every square meter of the earth's surface (Wm^{-2}), and they warm by about 25 Wm^{-2}. Recall that 60–100 W is the energy of a light bulb. The net effect of clouds is to cool the planet by about 25 Wm^{-2}. A change in cloud radiative effects of just 10 % would be 2.5 Wm^{-2}. Human forcing to date is about ~ 2 Wm^{-2} as detailed in IPCC. (2013). "Summary for Policymakers." In T. F. Stocker, et al., eds. *Climate Change 2013: The Physical Science Basis. Contribution of Working Group I to the Fifth Assessment Report of the Intergovernmental Panel on Climate Change.* Cambridge, UK: Cambridge University Press.

[6]For a summary of current state of knowledge, consult Boucher, O., et al. (2013). "Clouds and Aerosols." In T. F. Stocker, et al., eds. *Climate Change 2013: The Physical Science Basis. Contribution of Working Group I to the Fifth Assessment Report of the Intergovernmental Panel on Climate Change.* Cambridge, UK: Cambridge University Press.

5.7 Applications: Impacts of Tropical Cyclones

At the end of this and subsequent chapters we will present a section describing a particular application and use of climate models relevant to the topics of the chapter.

In the atmosphere, most of the impacts of climate change are felt through extreme events. Today's extreme events provide a framework for developing integrated planning for climate change. In particular, tropical cyclones are one of the most difficult types of extreme events to predict and understand in the atmosphere. Short-term weather forecasting is difficult enough. It is especially difficult to estimate the future climate characteristics of storms. Because the events are extreme and rare, we really do not have enough statistics to generate a reliable present-day distribution of storms. So storms often surprise us, such as in the example below.

In late October 2012, Hurricane Sandy made landfall centered on the New York and New Jersey coastline. The tropical storm merged with a mid-latitude cyclone, forming a large storm system with impacts across large portions of the northeastern United States and Canada. Characteristically, this merger of weather systems leads to wide-scale damage away from the coastline.

Hurricane Sandy received widespread attention for a number of reasons. The storm had a large geographic extent and affected a heavily populated region. The intensity of Sandy was not extreme (Category 3 of 5, winds of 115 mph or 185 km/h). The **storm surge** in New York City and northern New Jersey was record breaking, however. The resulting monetary damages in the United States were second only to Hurricane Katrina, which hit New Orleans in 2005. Sandy took an unusual path, with a sharp westward turn. This was well predicted a few days in advance by different forecast models. The forecast model of the European Center for Medium-range Forecasts predicted the storm track change two days earlier than the Global Forecast System of the U.S. National Weather Service. This discrepancy in forecast skill was widely reported, with public and policy consequences. Furthermore, there was much public discussion that the unusual path of the storm was consistent with hypotheses that the movement of weather systems is responding to global climate change. These factors helped to invigorate public discussion about climate change, including the discussion of science-based uncertainty, and a discussion of using models for prediction—in this case, weather forecast models.

The trajectory of Sandy at landfall was verifiably unusual, and it contributed to the large amount of damage associated with the storm. There was major environmental and infrastructure damage. There was significant environmental damage due to failures of sewage systems, making explicit the relationship of infrastructure, environment, and weather. There was extraordinary damage to transportation systems, residential structures, and commercial property. Because New York City is important to the global economy, economic impacts were amplified. The broad extent of the hurricane impacts revealed different levels of preparedness and response as well as relationships between policy and practice.

Much of the damage of the storm was related to the flooding, and specifically the storm surge of wind-driven seawater into New York City and towns along the New Jersey coastline. The region has seen sea-level rise on the order of 1.5 feet (0.4 m) since 1900. Some of this sea-level rise is attributable to climate change. Other factors include sinking of the land and variability of oceanic and atmospheric processes (see Sect. 6.7). In the next decades, the rate of sea-level rise is expected to increase; climate change will come to dominate the other causes of sea-level rise. The certainty of sea-level rise and the role of flooding due to hurricanes can help to manage planning for future tropical cyclones. Sandy is also a good example of how the different parts of the climate system (oceans and atmosphere) interact to create impacts.

There is no doubt that tropical storms will continue to have severe impacts on both the built and natural environment. There is no doubt that increasing sea levels will amplify the risks related to tropical storms. The risk due to sea-level rise is not limited only to tropical storms. Wintertime mid-latitude storms (called Nor'easters in the eastern United States) have some similar impacts to hurricanes. Hurricane Sandy demonstrated that the path of a geographically large, but not necessarily intense storm, can have large and costly impacts. Planning for increasing risk and frequency of storm-related flooding due to sea-level rise is warranted. Detailed scientific arguments about the frequency and intensity of tropical storms have only incremental impacts on most planning.

Sandy also exposes the importance of monitoring the emerging observational climate research to understand how it aligns with model projections. For example, two recent Pacific cyclones, in November 2013 (Typhoon Haiyan) and March 2015 (Cyclone Pam), had record winds at landfall, providing anecdotes of possible strength increases. The flooding of the island nation of Vanuatu by Cyclone Pam is viewed by many as a preview of a future with higher sea levels. The large diameter of Sandy and a number of other recent storms is consistent with the prediction that the geographic extent of storms is increasing.[7] In addition, evidence indicates that, in the Northern Hemisphere, tropical storms are having influence farther north than in the past.[8] Similarly, there is growing discussion that tropical storms are occurring earlier and later in the year,[9] consistent with a warming planet. There is additional risk if tropical storms are more likely to be present in areas where they were previously absent. These risks need to be taken into account by decision makers in their planning, and scientists need to focus on scientific investigations that can quantify and reduce the uncertainty in future projections of the characteristics of tropical cyclones.

---

[7]See   http://www.realclimate.org/index.php/archives/2015/03/severe-tropical-cyclone-pam-and-climate-change/.

[8]Kossin, J. P., Emanuel, K. A., & Vecchi, G. A. (2014). "The Poleward Migration of the Location of Tropical Cyclone Maximum Intensity." *Nature, 509*: 349–352.

[9]Kossin, J. P. (2008). "Is the North Atlantic Hurricane Season Getting Longer?" *Geophysical Research Letter*, L23705, doi:10.1029/2008GL036012.

5.8 Summary

Climate models of the atmosphere come in many shapes and sizes, ranging from simple idealized models with no dimensions, to complex models with high spatial resolution over particular regions or the whole planet. Both regional and global climate models are valuable. Regional models are often coupled to other scales to perform detailed experiments. The processes represented in an atmosphere model include the equations of motion and the various forcing terms that come from energy flows, motion, and transformations. Moisture is critical in the transformation and storage of energy.

Models have to approximate processes occurring on many scales with parameterizations to derive the source and sink terms for the equations of motion and heat that are iterated forward in time. Key processes include clouds, radiative flows of heat, and chemistry, as well as small-scale forcing of motions. Representing the variations in processes on small scales correctly can be difficult. Most of the uncertainty in the models results from missing or incorrect processes and particularly from the interactions across scales that cannot be represented. Feedbacks between processes can be complex and are not easy to observe. Cloud feedbacks are highly uncertain in atmosphere modeling. We return to many of these ideas in Chaps. 6 and 7, on the other model components of the climate system.

Key Points

- A hierarchy of atmosphere models exist.
- Global models reproduce the general circulation patterns of the climate.
- Physical processes in atmosphere models are complex. Clouds and radiation are key processes. Water is critical for moving heat in the atmosphere.
- Small-scale (subgrid) variations make parameterization difficult in atmosphere models.
- Feedbacks between processes are important, and clouds are the most uncertain.

Chapter 6
Simulating the Ocean and Sea Ice

The ocean covers about 70 % of the surface of the earth. Water is more dense than air: The top 33 ft (10 m) of ocean has the same mass as that of the entire atmosphere. As long as we remain near the surface of the earth (in the climate system), mass is equivalent to weight. In addition, the heat capacity of water (for the same mass) is four times larger than air. Thus, it takes more energy to raise the temperature of the ocean by the same amount for the top 33 ft (10 m) than for the entire atmosphere above it. Putting it another way: The top 10 m of ocean holds more energy than the atmosphere above it. Including the rest of the ocean below 10 m, the ocean is a much larger reservoir of heat than the atmosphere. Thus, the heat content of the ocean is a critical part of the climate system.

The ocean is also "stratified" with a series of shallow surface circulations and a deep ocean circulation (see Chap. 2). Large parts of the deep ocean do not interact rapidly with the ocean surface and hence can store heat away from the atmosphere. The ocean can serve as both a source and a **sink** of heat to the surface climate system (the atmosphere and land) on very different timescales from days to weeks up to hundreds of years. Thus, the ocean acts like a giant and slow reservoir that holds and redistributes energy in the climate system. It can also store carbon in several different forms, and that also makes it an important part of the carbon cycling through the climate system.

Simulating the ocean is critical for understanding climate on many scales. Some of the critical aspects are the exchange of heat and water with the surface, and the role of heat and **salinity** (the proportion of salt) in altering the density of the ocean. Density plays an important role in the ocean: Heavy water sinks; light water rises. The general nature of the **ocean circulation** cannot be understood without it. The basic elements of the ocean circulation are a result of the ocean boundaries (topography), the rotation of the earth, surface winds, and the changes to water density by changing the heat and salt content of water. These factors ultimately drive the ocean circulation, and they need to be represented in ocean models and properly coupled to the other parts of the climate system.

The cryosphere ("ice" sphere) contains land ice (ice sheets and glaciers), seasonal snow on land, and sea ice. Because land ice and snow are linked to land models, we discuss them in Chap. 7, on terrestrial systems. But it is logical to discuss models of sea ice in this chapter, as they are tightly coupled to the ocean. Sea ice is a critical

© The Author(s) 2016
A. Gettelman and R.B. Rood, *Demystifying Climate Models*,
Earth Systems Data and Models 2, DOI 10.1007/978-3-662-48959-8_6

part of the climate system because it strongly affects the surface albedo (white ice is much brighter than the dark, ice-free ocean, and it reflects more light back to space), and it also affects the surface energy coupling between the atmosphere and ocean (acting like an insulating blanket that allows the ocean to retain heat). Because of this, even though the cryosphere is a small area of the planet, it is an important part of the climate system, and it is critical at high latitudes.

In this chapter, we discuss ocean models and compare and contrast them with atmosphere models. We also cover models that simulate sea ice.

6.1   Understanding the Ocean

The key aspects of the ocean in the climate system can be described by under-standing the ocean structure, what drives or forces the ocean, and how this gives rise to the ocean circulation.[1] Modeling the ocean requires a representation of its structure. As with the atmosphere, the critical part is understanding what processes force the ocean, and then creating an appropriate representation of the physics of the forcing and structure that give rise to the ocean circulation.

6.1.1   Structure of the Ocean

Figure 6.1 is a simple schematic of the ocean. Think of it as a cross-section of one ocean basin, like the Atlantic from the South Pole to the North Pole. The ocean is divided vertically into a **mixed layer** near the surface (the top 50–100 m), where the ocean interacts rapidly with the atmosphere. Below the mixed-layer region, the effect of mixing with the surface gets smaller. Then deeper in the ocean (usually several hundred meters below the mixed layer) is a region of sharp gradients called the **thermocline** (*thermo* = heat and *cline* = gradient). Properly, this gradient is the **pycnocline** (*pycno* = density). The **surface ocean** lies above the pycnocline and contains the mixed layer. Below the pycnocline gradient, water is much colder and denser, and it exchanges very slowly with the surface. This is the **deep ocean** (or "abyss"), where the water temperature is nearly uniformly cold (pretty close to freezing). In high latitudes near the poles, the water is nearly the same temperature from the surface to the bottom, and there is usually not a thermally driven density gradient separating the surface and deep ocean. The density gradient in high lati-tudes is provided by changes in salinity (**halocline**; *halo* = salt).

The ocean in the tropics and mid-latitudes is stratified by density, with lighter water (usually warmer) on top and cold water beneath, illustrated by warm and cool

---

[1] A good detailed but qualitative treatment for the general reader is in Vallis, G. (2012). *Climate and the Oceans*. Princeton, NJ: Princeton University Press.

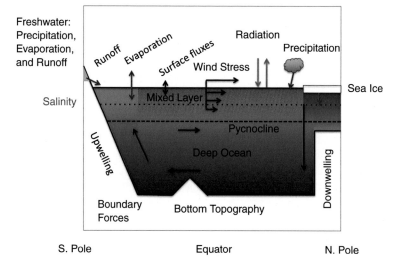

Fig. 6.1 Ocean schematic. This view of the ocean shows the mixed-layer boundary (*dotted line*) and the pycnocline or thermocline boundary (*dashed line*), with the deep ocean below. Changes of salinity are in *red*; changes in water are in *blue*; and evaporation, precipitation and runoff, and changes in energy (radiation) are in *green*. Ocean motions are in *black* and include wind stress and surface currents, and downwelling and upwelling in the deep ocean

colors in Fig. 6.1. At high latitudes, the temperature is cold all the way from the surface to the bottom (all cool colors). This means that, at high latitudes, surface water and deep water can easily mix if the density of one changes slightly (less dense water rises; more dense water sinks).

This structure of a mixed-layer, thermocline, and deep ocean arises from the interaction of the rest of the climate system with the ocean, and the properties of salty water. The density of water is a critical part of understanding the ocean. Ice is less dense than liquid, and it floats. In the same way, fresh water is less dense than salty water (salinity increases density), and warm water is less dense than cold water. Thus, when ocean water cools or acquires more salt, it gets more dense and, if denser than the water below it, sinks.

6.1.2   Forcing of the Ocean

The mixed layer is forced by **surface fluxes** (exchanges with the atmosphere). These exchanges include both exchanges of heat, momentum, and masses of salt and water. Heat enters the ocean at the surface by solar radiation filtering through the atmosphere. The exchange of heat, and the evaporation of water, changes ocean temperature. Changing temperature also changes the density of water. There are exchanges of freshwater between the ocean and atmosphere by precipitation and evaporation, and even from the land surface through rivers (see Fig. 6.1). Evaporation requires energy and removes heat from the ocean. Freshwater (without salt) also leaves the ocean when sea ice forms. Salt is expelled as sea ice forms. Freshwater returns to the ocean as sea ice melts (which might be a season or a year later). The addition or removal of water generally conserves salt content in the ocean, so as water is added or subtracted, the density changes. Atmospheric winds also push on the ocean, transferring momentum (often called "**stress**" on the ocean). The stress from wind on the surface generates surface ocean currents. These processes are indicated in Fig. 6.1. Note that almost all of these processes will affect the ocean circulation (cause water to move) either by pushing it (wind stress), or by changing heat or salinity that alters density (causing water to rise or sink).

Cold, dense surface waters at high latitudes form a link to the deep ocean, below the pycnocline. The deep ocean is driven by the density-driven motion of water, as well as by boundary forces on the side and bottom, and ocean topography (**baythemetry**) that alters circulations is similar to the way mountain ranges alter the atmospheric flow. Note that the ocean has boundaries on all sides and boundary layers on all sides. The atmosphere, however, has a boundary layer only at the bottom, with basically no mass at the top.

6.2   "Limited" Ocean Models

Like the atmosphere, simulating the ocean dynamically with a finite element model is a difficult task, requiring lots of approximations. As with the atmosphere, there are several types of ocean models. We concern ourselves mostly with ocean general circulation models (GCMs), which are similar to GCMs for the atmosphere. Ocean GCMs have complex representations of the ocean circulation throughout its depth, full ocean topography, and parameterizations for small-scale mixing processes at the surface and within the ocean.

Beyond the full representation of the ocean structure in Fig. 6.1 are simpler types of ocean models. These models try to represent individual ocean basins or just regions of those ocean basins (Fig. 6.2). As with models of the atmosphere, they are often called regional models. Basin-scale regional models may be used for experiments that look at coupling between the ocean and the atmosphere. Regional ocean models are similar to limited area atmosphere models discussed in Chap. 5: They are often used at high resolution, coupled to limited area atmosphere models to represent details of regional weather or climate. They are forced at their boundaries by observations or by output from other models. These regional models may not include a deep ocean circulation. The deep ocean circulation is often specified in regional ocean models. Regional ocean models are designed primarily to represent the region above the thermocline that changes relatively quickly and responds to the surface climate system. These models are focused on the surface properties of the ocean that interact on fast timescales (days to weeks or seasons) with the atmosphere, ice, and land surface. They have detailed representations of surface

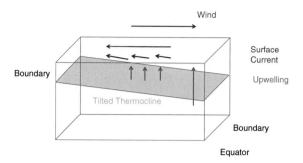

Fig. 6.2 Ocean basin. Representation of an ocean basin from the equator to a northern boundary. Trade winds near the equator (*red*) cause water to move along and away from the equator, also tilting the thermocline or pycnocline. The motion of water away from the equator causes upwelling along the equator

exchanges, radiation, and wind stress that forces the mixed layer of the upper ocean. They may have detailed representations of the ocean floor topography, often near coastlines.

A commonly used type of global ocean model seeks only to represent surface processes, and the fast communication of the ocean with the rest of the climate system. These are global **mixed-layer ocean models** (sometimes referred to as "slab" ocean models). Mixed-layer ocean models are global in scope, representing all ocean basins, but they are focused on the upper ocean, not the deep ocean, that is, the region above the pycnocline in the mixed layer in Fig. 6.1.

The advantage of mixed-layer ocean models for simulating the climate system is that they contain smaller amounts of water mass (from 10 to several hundred meters deep). Like the limited area models described earlier, these models have a limited area, but here the limit is in depth. The models typically have no circulation but are simply an energy balance equation: The temperature of a fixed-depth layer is determined by the heat and water coming in and out at the surface, and the specified heat from a bottom boundary. They do not include currents or water motions.

A mixed-layer ocean model just adjusts the temperature of the slab of water to respond to the surface energy budget from the atmosphere, and an assumed interaction with the deep ocean below. Mixed-layer ocean models need to have a bottom boundary condition, usually a specified movement of heat to and from the unresolved deep ocean below. Like a limited area model in the horizontal, the boundary conditions often come from observations or from a more detailed ocean GCM. The models are designed to reproduce the ocean surface temperature and interact with the rest of the climate system given a specified ocean circulation pattern below. The advantage of this configuration is that it reduces the heat capacity of the ocean and allows it to come into balance much faster: decades rather than centuries.

The disadvantage of mixed-layer ocean models is that they require a fixed assumption about the ocean circulation, so if the climate is too different from what is assumed by the specified energy transfer from the deep ocean, the ocean model may not yield a reasonable result. Because water cannot move horizontally in this class of models, if too much heat is removed or added to a grid box in the ocean, the bottom heat exchange is fixed and the temperature of the surface temperature may change a lot. In a full ocean model, however, a change in surface heat exchange may change density and cause water to move. The motion of water can take heat away from the surface and alter the heat exchange with the deep ocean, which is fixed in mixed-layer models. So mixed layer ocean models need to be used with some caution, and for more reliable climate system calculations, full models with the ocean circulation are used.

6.3   Ocean General Circulation Models

We divide our discussion of ocean modeling into a description of the grids and dynamics used in an ocean model, the deep ocean and the thermocline, and the surface ocean and the mixed layer. The surface ocean and the mixed layer are of primary concern to many of the simplified models described above as well.

6.3.1   Topography and Grids

As with the atmosphere, GCMs of the ocean are finite element models. However, whereas atmosphere models need be concerned only with a bottom boundary, and energy input at the top, the ocean has top, bottom, and side boundaries. The bottom boundary is similar to the topography we are familiar with on land: The ocean has mountains and valleys and complex topography that affects the circulation. The ocean has a top boundary that is critical for coupling to the bottom boundary of the atmosphere, and the primary forcing of the ocean occurs here. The horizontal boundaries of the ocean basins mean that the ocean grid is not global: Not every point needs to be represented, since some latitudes and longitudes have no ocean.

Note that these grids may change over time if the sea level changes. During the last ice age, when the sea level was about 330 ft (100 m) lower, there were significant differences in topography (bathymetry): The Bering Strait between Siberia and Alaska was a land bridge (enabling *Homo sapiens* to walk to the Americas from Asia), and the region between Indonesia and Australia was mostly land as well (with one or two channels).

Because the ocean does not occupy the whole planet, different grids have been constructed. First, a latitude-longitude grid is possible, but with boundaries and cells that do not exist because they are land points. However, having grids that converge into a single pole creates problems because the same number of cells exists at all longitudes. For a 1° longitude there are 360 points at the equator, and each point represents 68 miles (110 km). But at the poles, the distance around the earth goes to zero, so the size of the points becomes small. At 80° latitude, each degree of longitude is only 12 miles or 20 km. This introduces computational problems. Wind or currents may move air or water farther than one grid box in a single time step. For many processes, this creates problems in accounting for all the energy and mass. In the Southern Hemisphere, the pole and regions around it are conveniently on land (Antarctica) and the South Pole has no ocean, so this problem is not as acute. But the Arctic Ocean is difficult to represent on a traditional latitude-longitude grid.

Two approaches can get around this problem of having a pole in the ocean. "Equal area" grids are possible, where faces of a cube are projected onto the sphere so that most grid points have similar area. Another method is to "shift" the pole of an ocean grid onto land. The mathematics is complex, but this approach is computationally efficient and avoids mathematical problems with very small grid boxes (wedges) at the poles.

Another type of grid is not regular, and thus is often termed "unstructured." These **unstructured grids** often have finer resolution in critical areas for either global or regional simulation. The goal is to increase resolution where it matters, for example, in regions with narrow straits or important bottom topography, to better represent the boundaries of the ocean basins. This puts resolution where it is needed and is more efficient than having high resolution everywhere. Because ocean topography is such an important forcing term for ocean circulations (more so than in most regions of the atmosphere), these variable resolution grids are more common in ocean models.

6.3.2   Deep Ocean

The deep ocean (below the mixed layer) has a global circulation called the **meridional overturning circulation** (Fig. 6.3).[2] The circulation has one component driven by **buoyancy**, where water sinks because it gets heavier. Colder water is heavier and saltier water is heavier, so it convects when it sits on top of lighter water. **Convection** is the same buoyancy-driven force in the atmosphere, where lighter (warmer) air lies below heavier air, and it rises, forming clouds. The buoyancy component of the ocean overturning circulation is called the **thermohaline circulation**, driven by heat (*thermo-*) and salt (*-haline*). Surface winds also help drive the circulation. The ocean circulation is regulated by ocean topography that controls where water flows. Both processes interact to produce the deep ocean circulation: The density-driven thermohaline circulation acts like a "heat engine," whereas the components driven by surface winds might be more analogous to a "pump." It is the combination of these forces that results in the deep ocean circulation.

---

[2]Figure 6.3 is based on a figure from Rahmstorf, S. (2002). "Ocean Circulation and Climate During the Past 120,000 Years." *Nature, 419*(6903): 207–214. doi:10.1038/nature01090. This paper is also a good introduction to the ocean circulation.

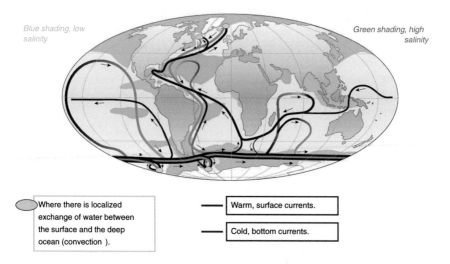

Fig. 6.3 Deep ocean circulation. The schematic shows the warm surface currents and deep bottom currents. *Orange* regions indicate where deep water forms. *Blue shading* indicates low salinity, and *green shading*, high salinity. Deep water forms in the North Atlantic and near Antarctica and flows throughout the ocean basins. Figure adapted from Rahmstorf (2002)

Representation of the ocean circulation in models is dependent on a number of factors. Of prime importance are the processes regulating the changes in density driven by salinity changes and temperature changes. In polar regions, sea ice forms (expelling salt into the ocean) and the ocean cools by losing heat to the relatively colder atmosphere. Both cause increases in density, and the surface water becomes denser than the water beneath. This water sinks to the bottom of the ocean, forming **bottom water**. Bottom water formation happens in both the Arctic and the Antarctic in the orange regions in Fig. 6.3. Antarctic bottom water forms just off the Antarctic ice shelf and is generally denser due to very cold temperatures, flowing throughout the bottom of the earth's oceans, constrained by bottom topography (see Fig. 6.3). Because masses of water do not mix very well, the ocean water keeps its properties (heat, salt, and trace chemical distribution) for long periods of time. Water is often named for, or characterized by, where it last encountered the surface. The oceans are cold at the bottom because the water comes from high latitudes. There is little warmth from the seabed. Water is warm when it is warmed at the surface by the sun.

The bottom water flows equatorward in the Atlantic (south). In the Pacific and Indian oceans it flows equatorward from Antarctica (north), starting a global circulation (see Fig. 6.3). The newly formed bottom or deep water may not see the surface again for a thousand years, and it will largely preserve its characteristics of salt and heat throughout that time, changing only slowly. Ocean circulation speed can be observed by looking at chemicals[3] in deep ocean water: Long-lived and inert industrial chemicals from human origins in the air mix into seawater in trace amounts and can be seen spreading out with the deep water in ocean basins, such as from the North Atlantic into the tropical Atlantic. The chemicals are present in microscopic quantities, and are inert and not dangerous. The presence of these industrial compounds provides a record of when the water last was at the surface.

One of the key processes to correctly represent the buoyancy-driven circulation is the vertical mixing of water masses as the denser (colder, saltier) water sinks. This is a density-driven convective process, physically the same type of process that occurs during atmospheric convective motions that drive deep thunderstorm clouds. Simulating the mixing that occurs during convection is important for representing the resulting composition (heat, salt) that the bottom water has. Typically the process is represented either as a simple adjustment necessary to get the water column "stable" again, with heavier water on the bottom, or as a "diffusion" process that depends on the large-scale vertical density gradient. Ocean vertical mixing is a critical parameterization in which the representation of a fundamental physical process (density) must describe complex interactions in space and time that are often below the resolution of the model.

In general, ocean models can do a good job of simulating the thermohaline circulation patterns, which are governed by the large-scale position of the continents. However, the amount of deep water that forms and, hence, the "speed" or mass in the overall circulation can vary quite a bit. The overall mass transport and speed of the circulation is dependent on a balance of processes in an ocean model: formation of deep water by density changes in the Arctic and the Antarctic, wind-driven upwelling around Antarctica, bottom topography and forces throughout the ocean, and diffusion of heat in the interior of the ocean. Surface stress from wind alters surface currents. The surface stress of winds around Antarctica pushes water offshore, and creates cold conditions where sea ice forms, also creating cold and salty dense water that forms the Antarctic bottom water. Finally, the surface return flows, such as the Gulf Stream (see below), transport their mass in eddies and a mean circulation. The flows are created by the rotation of the earth acting on ocean water in confined basins.

The deep ocean circulation exists because of density differences and the tendency for stratification. The global circulation meanders through ocean basins as a result of topography. It is affected by the surface exchanges of heat and salinity. It is also affected by surface-driven forcing against boundaries that cause upwelling. We consider these complex interactions and their representation when we discuss the surface ocean (Sect. 6.3.4).

The ocean circulation is forced at large climate scales, but there is also variability of density, temperature, and topography at small scales. These small-scale differences lead to responses in the circulation. Some of these responses are very large

---

[3]Chlorofluorocarbons, or CFCs. The same chemicals that deplete the stratospheric ozone layer are inert in the absence of sunlight (in the ocean).

scale and depend mostly on the fundamental equations of motion and energy transfer, making them easier to represent in ocean models, but others are small scale and give rise to small-scale motions (**eddies**) throughout the ocean. These eddies may have important consequences for mixing ocean water properties that can ultimately affect the density and the circulation.

6.3.3 Eddies in the Ocean

Many of the critical processes that force the ocean, and the most uncertain ones, depend on variations at the small scale. For simulating the ocean, one of the most important problems is that internal wave motions and eddies travel slower because of smaller density gradients. This means that they maintain their coherence on smaller scales. The small size also means that current variations (ocean "weather systems") have a space scale that is smaller than the same critical scale in the atmosphere, and a longer timescale. The major oceanic flows, driven by the large-scale forcing of the wind, boundaries, and density variations of upwelling and downwelling, carry most of their energy in meandering small-scale eddies. This is a bit like the individual storm systems in the atmosphere that move large amounts of water at mid-latitudes.

Figure 6.4 shows ocean surface temperature variations in the **Gulf Stream** off the east coast of North America.[4] The Gulf Stream provides a vivid example of the complexity of ocean currents and eddies. The warm flow poleward is not a straight "highway" but meanders with swirls relating to instability on the sides of the current and interactions with the bottom topography. An idealized current might be like a drainage ditch with concrete sides (e.g., uniform lines in Fig. 6.2), whereas the actual ocean currents have lots of small-scale eddies like an uneven meandering stream (see Fig. 6.4). Eddies carry more of the energy in the ocean than they do in the atmosphere. As a result, ocean models are often run on finer grids than atmosphere models. A typical atmosphere model is 100–200-km resolution, and high resolution is 25 km, whereas a high-resolution ocean model would be 10 km, and standard resolution 25–100 km. Even so, these eddies are still often much finer scale than the grids in an ocean model, and representing how they form and evolve, and their effect on the flow, is a central problem of ocean modeling.

Ocean models contain several different types of eddies. As computation enables finer resolution, more of the mesoscale at 6–62 miles (10–100 km) can be resolved. But this means that representations of the sub-mesoscale from 0.6–6 miles (1–10 km) become more important. Because the motions are slower, and mix less than the atmosphere, the scales of motion in space become finer. More of the energy in the flow is contained in these structures. They are present throughout the depth of

---

[4]The image is of the surface temperature of the ocean from the Atmospheric Infrared Sounder (AIRS) satellite instrument. Public domain image credit: NASA.

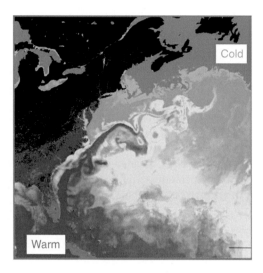

Fig. 6.4 Gulf stream. Satellite surface skin temperature of the North Atlantic from the atmospheric infrared sounder (AIRS). Coldest water in *blue*, warmest water in *red*; *orange*, *yellow*, and *green* are in between. The "Gulf Stream" of warm water from the Gulf of Mexico into the N. Atlantic is clearly seen with all of the eddies around it. *Image credit* National Aeronautics and Space Administration.

the ocean, but we usually see them only in the surface layer (such as in Fig. 6.4). Yet the large-scale effects of eddies mixing water are crucial for the largest climate scales (such as the flow in the Gulf Stream in Fig. 6.4). It is as if the large scales (think of the highway or the concrete drainage ditch) require the small scales to handle the flow (or energy) of the circulation. Currently, a great deal of ocean model development is focused on consistent representations of these eddies and their effects on large-scale mixing and overall circulation. It also drives ocean models to finer resolutions (6 miles or 10 km or even finer). Similar scale problems exist in the atmosphere with scales of motion that are within an order of magnitude or so of the grid scale: too coarse to resolve properly, but too fine to represent statistically.

6.3.4   Surface Ocean

The **surface ocean** is the primary region where the ocean communicates with the climate system. It is forced primarily by exchanges of energy, water, and momentum across the top boundary of the ocean. Whereas the deep ocean is driven by vertical gradients and vertical mixing, the surface ocean is mostly driven and affected by wind-driven forcing combined with topography (boundaries) and the effects of rotation. The forcing is highly variable and gives rise to eddies that comprise the large-scale flows. As evident in Fig. 6.4 and discussed earlier, representing the effects of these eddies in the ocean is a central problem of ocean modeling. These eddies are present in the wind-driven **gyres** (circular currents) we describe later in this section, and instabilities in the surface forcing can generate eddies. Furthermore, when water at the surface is pushed away from boundaries, it can cause upwelling of water from the deep ocean.

The surface ocean is forced by variations in atmospheric wind patterns. As with water or air that vertically wants to achieve its "correct" place in the column of density (stratification), atmospheric winds blow from high pressure to low pressure. At the surface, these pressures are coupled to patterns of energy input (which creates warmer or colder temperatures) and evaporation of water from the surface (which also can cool temperatures). Pushing water around an ocean basin causes currents, and it also causes slight differences in the height of the ocean as water "piles up" when forced against a boundary. The changes are too small to see, just a foot or less (10–25 cm) over thousands of miles, but stacking-up water means water wants to flow down this gradient. These forces act slowly and are affected by the earth's rotation. The good news is that the description of these forces can be well represented in equations in large-scale models. It is not a problem that the effect is small: It is representable because the effect occurs on a large scale. The ocean, of course, plays a large role in surface temperature and evaporation; hence, the atmosphere and ocean circulations are tightly coupled at the surface. This is true in many regions of the planet, from the tropics to the Arctic and the Antarctic.

From the perspective of the ocean, the wind induces a force on the ocean called a stress. **Wind stress** is visible in surface waves: The stronger the wind, the bigger the waves. This pushes the surface water, and the force is communicated through the body of the water column for some depth (decreasing as one gets deeper). The direction also changes with depth due to friction. The wind stress, combined with the boundaries in the ocean and the rotation of the earth, results in what is called **Sverdrup balance** (after a Norwegian oceanographer) between the force of the wind and the north-south transport of water. Water piles up on the west side of ocean basins in the tropics, and flows poleward, with an equatorward flow on the eastern side.

There is another feature of motion, however, and that is induced by rotation of the earth. Because the earth is rotating, there is an apparent sideways force, the **Coriolis force**, pushing to the right of motion in the Northern Hemisphere, and to the left of motion in the Southern Hemisphere.[5] As water flows to the west at the equator, this creates poleward motion. Thus, water comes from below along the equator to replace it (see Fig. 6.5). For water flow along a coast, the same force occurs. Equatorward return flow off the western part of continents (western North America and South America, and the Atlantic coast of Africa and Europe) induces

---

[5]Why is it in opposite directions when the earth spins the same way? The reason has to do with the angle of motion relative to the axis of the earth's rotation. For a complete description, see https://en.wikipedia.org/wiki/Coriolis_effect.

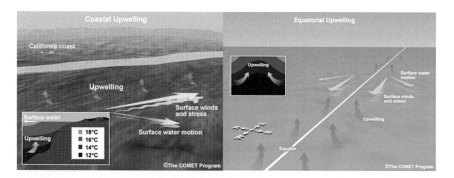

Fig. 6.5 Surface ocean. Coastal upwelling (*left panel*) due to southward surface winds along the coast, bringing up cold water. Equatorial upwelling (*right panel*), surface winds along the equator to the west (easterlies), causing water to move away from the equator and resulting in upwelling along the equator. Figure from the COMET program

offshore flow, also causing upwelling water from below to replace the surface water moving offshore. The Gulf Stream (see Fig. 6.4) is a manifestation of the poleward flow (the Kyushu current off of Japan is the Pacific version). So the flow induced by the large-scale rotation of the earth and the wind stress then has additional components from the basic physics of fluid on a rotating sphere.

The forces describe something called **geostrophic balance**: Water that is moving is affected by the earth's rotation, causing additional water motion. The oceanic surface gyres are driven by wind and the Coriolis force: The tropical westward flow of water in each hemisphere results from wind forcing along the equator (the trade winds). This induces a poleward Coriolis force that occurs on the west side of ocean basins. The eastward flow of water at mid-latitudes (again driven by prevailing winds) then induces an equatorward flow on the east side of basins.

The friction of ocean water combined with the Coriolis force also means that the force on ocean water is not in the same direction as the wind forcing (stress). If you push ocean water with wind, it tends to move a bit to the right of the direction of force (dictated by the earth's rotation). Because it is frictional, the layer below moves a bit more to the right, so that the net water motion is almost at a right angle to the wind stress. Along coastlines and the equator, this induces upwelling (illustrated in Fig. 6.5) from below as water flows away from a coast, or away from the equator. The cold upwelling water creates some of the temperature patterns in the surface ocean.

The mixing of momentum and heat down from the surface into the mixed layer is an important process for representing the structure of the ocean. Small-scale density gradients are induced by variations in temperature and salinity that make up this boundary layer at the top of the ocean. The ocean mixed layer is analogous to the atmospheric boundary layer above. Representing these fine-scale boundary layer processes in the ocean is important for coupling with the atmosphere above, not just for ocean circulations.

Exchanges of energy and water mass between the atmosphere and the surface of the ocean are also important. The ocean absorbs solar radiation, and some downward longwave (infrared) radiation from the atmosphere, and emits longwave radiation back. These exchanges help determine the temperature of the surface ocean. The exchange of freshwater with the surface is also a critical process. Water comes into the ocean from precipitation, rivers, or melting ice, and leaves by evaporation from the surface and formation of sea ice. Input and output water is fresh, so the addition or subtraction of fresh water with the same mass of salt will change the salinity and the density. Evaporation also is a cooling process, adding heat to the evaporated water, and removing it from the ocean. All of these exchanges of heat and mass of both water and salt must be accounted for exactly in an ocean model, and transferred to and from the atmosphere as appropriate. This is essentially a giant budget exercise. Like financial budgets track dollars, the surface energy and mass budgets have to track energy (watts) or mass (kilograms). The accounting has to be absolute: Even small systematic errors in mass and energy will be significant over long timescales.

6.3.5   Structure of an Ocean Model

The structure of an ocean model is internally similar to that of an atmosphere model. The ocean is divided into different grid cells distributed throughout the ocean. The ocean grids are often irregular and do not include points only on land. This makes them more complex. But ocean models at the same resolution have fewer points than atmosphere models, as they cover only 70 % of the earth's surface.

Ocean models also have a basic time-step loop. Typically, surface forcing from the atmosphere is calculated. This enables an estimate of the change in forcing on the ocean surface, and the change in pressure that will affect the height of the sea surface. When the atmospheric pressure drops, the ocean will tend to rise underneath it, and that water comes from somewhere else: inducing currents that need to be estimated.

Next, the different forcing terms on the ocean model, arising from different forces and parameterizations, are estimated. These include important parameterizations of eddies and eddy mixing. Changes to the mass of water and salt are estimated. One major difference between the atmosphere and the ocean is that, in the atmosphere, the forcing terms (clouds, radiative transfer) all occur independently in each column of the atmosphere: in one dimension in the vertical. In the ocean, the eddies mix horizontally as well as vertically, in three dimensions.

Finally, all these forcing terms for heat, currents, and even salinity are applied with the equations of motion for fluid (water) on a rotating sphere to get the resulting motions of water and changes to density in each grid location in the ocean. The tracers for chemicals in the ocean are updated. Then the revised state is iterated forward in time and the process begins again. This is very similar to how an atmosphere model works as described in Chap. 5.

6.3.6   Ocean Versus Atmosphere Models

There are many similarities between ocean and atmosphere models. Ocean models use most of the same scientific principles for fluid motion on a rotating sphere that apply to the atmosphere. There are difficulties in representing important mixing and transformation processes at small scales. Minor constituents (salt in the ocean, water in the atmosphere) play a major role in the general circulation. The grids and computational techniques of finite element modeling, and the use of subgrid-scale parameterizations in ocean models, are similar to those found in atmosphere models. But there are also major differences between the ocean and the atmosphere. The ocean has boundary layers on all surfaces (not just at the interface with the atmosphere). Ocean models have complex topography on both the bottom and the sides, and the ocean is effectively divided into basins (the five ocean basins of the Atlantic, Pacific, Arctic, Indian, and Southern Oceans). The ocean has more energy in the eddies and at smaller scales than the atmosphere, making their representation critical. Ocean models are forced by the atmosphere above, and that forcing is transmitted throughout the depth of the ocean, giving rise to eddies, surface currents, and the deep ocean circulation.

Thus far, we have focused on the physical description of the ocean. We return to some of the biogeochemical cycles in the ocean in Sect. 6.5.

6.4   Sea-Ice Modeling

Sea ice[6] is coupled closely with the ocean circulation. It concerns a representation of the freezing and melting process of ice and snow on top of the sea ice. Sea-ice models now also treat the dynamic motion of ice. Salt in (or expelled from) the ice is important for the ocean, and the ice strongly affects the flow of energy between the atmosphere and ocean. Sea ice forms in unique conditions at high latitudes where the temperature is cold. Ocean water freezes at about 28 °F (−2 °C) because of its salt content. Since temperatures vary strongly over the year, there is a large annual cycle in the extent of sea ice. In much of the polar regions, the sun has more of an annual cycle than a daily one, and above the Arctic and Antarctic circles, there are long periods when the sun is always present ("midnight sun") or always absent ("polar night").

---

[6]For a review of sea ice in the climate system, see Marshall, S. J. (2011). The *Cryosphere*. Princeton, NJ: Princeton University Press, Chap. 5.

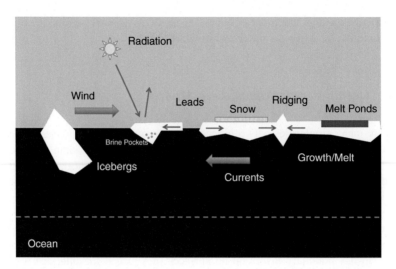

Fig. 6.6 Sea ice. Key processes in a sea ice model at the surface include winds, radiation, open areas between ice (leads), snow on ice and melt ponds. Key process at the base include currents, growth, melt and brine pockets

Sea-ice models must represent ice growth and melt. A schematic of many of the processes described is shown in Fig. 6.6. Models must account for the salt in seawater that is expelled back into the ocean, because this salt is critical for altering the density of the ocean at high latitudes. Along with temperature, it drives the formation of deep water at high latitudes. The salt content is dependent on the conditions of formation. Calmer water means more regular ice growth, and more salt expelled. In rougher conditions, **brine pockets** can become trapped in ice. Models must account for this inherently small-scale process to properly represent the exchange of salt and the density of the water underneath the ice.

Sea ice damps heat and moisture exchanges between ocean and atmosphere. This occurs because of the albedo contrast between bright ice and dark ocean and because ice is good at insulating the ocean from the atmosphere, reducing the surface exchange of heat, which must conduct through the ice, rather than convect (with density-driven motion) in the atmosphere. Convection is a lot more rapid than conduction (at least between ice, air, and water). So ice models must represent the conduction of heat. The heat conduction is dependent on the ice structure and can be altered by things like brine pockets and by the presence of snow on the ice. There is also a coupling of growth rate and thickness: Thin ice grows (and melts) faster because there is less resistance to conduction of heat through the ice to the atmosphere. Note that sea ice typically grows from the bottom, and the bottom may be irregular. In addition, snow falls onto the ice from the atmosphere, adding a little to the thickness. These are complicated processes in the thermodynamics of the evolution of sea ice.

In addition to the changes of state associated with ice formation and melting, the energy budget of sea ice must be accounted for. Radiation from the atmosphere, solar radiation (only in part of the year), and longwave radiation (all year) hit the ice. Ice and snow reflect some shortwave (solar) radiation, and absorb and emit longwave radiation. Sea-ice models must account for the radiative transfer in the top layers of ice and snow on the ice. As noted, the presence of snow on top of ice is quite important, as it changes the surface properties for both the absorption and scattering of light, as well as the conduction of heat (snow can "insulate" ice). The sea ice is a bit like the middle layer of a sandwich between the atmosphere and ocean, but it is critical for regulating exchanges between the two, and these exchanges have critical importance for how heat flows through the climate system.

These are thermal considerations, related to the conduction of heat through ice, its absorption, and the flux of heat to and from the atmosphere and ocean. The other critical aspect of sea-ice simulation is the dynamics, or motion, of sea ice. Sea ice is in constant motion, pushed by winds and by currents. The motion causes stresses in the ice and can cause it to deform. Sea ice is nearly flat, but with different layers that receive body forces from the ocean (bottom) and atmosphere (top). Thus, models must represent the momentum balance of the ice, the distribution of ice thickness, and the physics of the flow of ice in response to stresses (**rheology**). The thickness varies on small scales and is usually treated as a "thickness distribution" in any large-grid cell. The motion of the ice is predicted in response to the environment, stresses, and the internal structure of ice. Sea-ice motion can result in **leads** (open spaces) where there is divergent motion, and **ridging** in regions of convergent motion (see Fig. 6.6). Leads expose open water and often promote ice growth (or melt). Ridging increases ice thickness and helps ice survival: Thicker ice lasts longer. These processes are starting to be represented in sea-ice models. Many of these processes act on subgrid scales, occurring only in part of a grid box.

Sea-ice models act at the interface between the atmosphere and the ocean, and are subject to strong forcing from both. Traditionally they are often strongly coupled to ocean models and usually share the same grid as an ocean model. These ocean grids usually do not have a convergent "pole" in polar ocean regions, so the grid box sizes are usually nearly equal area.

As with the ocean, there are simplified sea-ice models. **Simplified sea-ice models** can assume a fixed distribution of sea ice (usually by month). More common is the use of thermodynamic considerations to estimate thickness and the local energy budget and energy fluxes. Mixed-layer ocean models are usually run with thermodynamic sea-ice models, since there are no ocean currents to move sea ice. A full sea-ice model adds a dynamic motion component to the simulation.

6.5   The Ocean Carbon Cycle

A new frontier of ocean modeling is the representation of the global cycle of carbon.[7] Carbon is a unique constituent of the climate system, since it passes through many of the different components: carbon dioxide (CO_2) as a gas in the atmosphere, into the land surface (as carbon-containing organic and inorganic matter), and also into and out of the ocean. CO_2 is dissolved in the water column, in chemical equilibrium with the atmospheric CO_2 pressure and ocean temperature. This enables ecosystems of aquatic plants to build biological matter (i.e., their bodies) with it, and forms the basis of the oceanic food chain. Carbon is present in calcium carbonate, which is also dissolved in water, and forms the shells of many marine organisms. Thus, when marine animals die, there is a steady buildup of carbon-rich sediments in the ocean. These oceanic carbon cycle processes are fundamental for affecting climate on long glacial and geologic timescales. They do not react very quickly (on timescales less than a century), but they may be important for understanding how ice ages occur, and how temperature and CO_2 vary with each other. The ocean is a vast store of carbon, and changing temperature and circulation may allow more or less carbon into the atmosphere, with resulting impacts on warming.

Models of the carbon cycle are beginning to be coupled with ocean and climate system models. They must represent different transformation processes for carbon based on organisms (biological carbon) and fundamental chemical processes (such as how much CO_2 is dissolved in seawater, which is a function of temperature). Some of these representations reflect simple chemical laws for how much CO_2 or carbonate is dissolved in seawater at a given temperature, and some are representations of biological processes. We treat the carbon cycle more fully in Chap. 7, on terrestrial systems.

6.6   Challenges

Ocean and sea-ice modeling is complex, and the different scales of motion (with small space scales, and very long timescales) pose a challenge for modeling. So, too, do the now-rapid changes in sea ice in the Northern Hemisphere. Some of these challenges are similar to the challenges faced by atmosphere models (such as the small scale of processes, and variability in a grid box), and some are unique to the ocean and ice.

---

[7]Archer, D. (2010). *The Global Carbon Cycle*. Princeton, NJ: Princeton University Press.

6.6.1 Challenges in Ocean Modeling

One major challenge in ocean modeling is a dearth of observations, particularly below the surface. The ocean is more difficult to observe than the atmosphere because it rapidly absorbs most wavelengths of radiation, and it is sparsely populated (by humans at least). So there are a lot less data beyond just the surface of the ocean. Most deep ocean data still have to be taken directly, which is not easy in a 4,000-m deep ocean. The rise of autonomous devices (buoys and small, unmanned undersea vehicles) enables remote measurements where unmanned submersible devices can rise and sink down to at least 2,000 m, recording temperature, salinity, and current measurements not unlike weather balloons in the atmosphere. Some have large fixed buoys at the top, and some drift, relaying their information to satellites when they surface. These systems are rapidly improving ocean observations and contributing to evaluation of ocean models, but observations of the deep ocean (below 2,000 m depth) are still very limited.

Another challenge in ocean modeling is properly representing the effect of small-scale eddies that cannot be explicitly simulated by a large-scale ocean model. Eddies move a lot of mass in the flow in the ocean, more like a meandering stream than a straight channel. This can be seen in the picture of small eddies in the Atlantic Gulf Stream in Fig. 6.4. Because the scale of eddies (6–30 miles, 10–50 km) starts approaching the ocean-model grid scale, it is difficult to represent them properly. Trying to represent a curvy flow in a stream is difficult if there are only one or two values for the current. One solution is "high-resolution" (6-mile or 10-km spacing) ocean models that "permit" the formation of eddies but are too coarse to resolve them properly.

Furthermore, ocean models have long adjustment timescales because of the deep ocean circulation. The use of simplified models of the mixed layer has come about since a full dynamic ocean model with a thermocline and a deep ocean circulation will reach a steady state (no change to climate with no external forcing) in about the time it takes for the ocean water to recycle, which is thousands of years of simulation. However, a mixed-layer model can reach equilibrium in only decades. The implication of this long timescale means that perturbations to the earth system will take thousands of years to equilibrate because of the slow processes in the ocean.

6.6.2 Challenges in Sea Ice Modeling

Sea-ice models have had quite a bit of recent success in simulating the observed distribution of sea ice (see Chap. 11). The comparison is complicated by the lack of observations of ice thickness. Since the arrival of full global weather satellite coverage in the 1970s, it is relatively easy to observe sea-ice coverage, but thickness observations remain elusive. The realism of ice thickness represented by sea-ice models is then uncertain. An additional complication is that the sea-ice distribution in sensitive seasons (like summer and fall) has been declining rapidly in the Arctic over the observational record, and it looks like the decline is getting faster (in terms of area of Arctic sea ice at a particular time, usually September). Most sea-ice models forced with atmospheric observations are able to reproduce the decline currently being experienced in the Arctic, but not necessarily the magnitude of the decline (they indicate less decline and more stable ice). Fully coupled models with an interactive ocean and atmosphere have a hard time reproducing the rapid decline of Arctic sea ice, likely due to the uncertainty in the forcing on the system going into these models and the complex interactions among ocean, ice, and atmosphere. It may be that the rapid sea-ice loss in the Arctic is a consequence of a long-term greenhouse warming signal, with short time period (i.e., a season or a year) additional variability. Because the sea ice is prone to melt and lasts from year to year, a set of events promoting loss in one year may result in lower ice the next year, and drive positive feedbacks. These surface feedbacks are discussed in Chap. 7.

6.7 Applications: Sea-Level Rise, Norfolk, Virginia

This case study demonstrates the roles of many stakeholders and the complex relationship between climate change science and other sources of information. Metropolitan Norfolk, Virginia, is a low-lying collection of cities and natural regions on the eastern coast of the United States. In addition to residential, recreational, and commercial activities, there is a large military presence, especially the U.S. Navy. Many private companies are associated with the naval presence, including unique dry-dock maintenance facilities.

Since 1971, flooding in Norfolk has increased from about 20 h a year to 130 h per year.[8] Between 1930 and 1997, only six storms brought **storm surges** (rising seas like a high tide due to wind and low atmospheric pressure) greater than 3 ft (1 m). Since 1997, there have been seven storms with surges greater than 3 ft. Though this part of the U.S. coast is often associated with hurricanes and tropical storms, wintertime Nor'easters are of equal importance when considering storm surges. This rapid increase in coastal flooding has sensitized the region to changes in sea level.

Analysis of the sea-level rise reveals that only part of it is due to the warming of the ocean and the melting of ice sheets and glaciers. The local land is sinking, partly because of rapid pumping of groundwater for residential and industrial use. In the past 100 years, the sinking land has been a larger effect than sea-level rise due to climate change. In addition to sea-level rise associated with the global average, there are strong local effects. These local effects are largely related to the variations

---

[8]Ezer, T., & Atkinson, Larry P. (2014). "Accelerated Flooding Along the U.S. East Coast: On the Impact of Sea-Level Rise, Tides, Storms, the Gulf Stream, and the North Atlantic Oscillations." *Earth's Future*, 2: 362–382. doi:10.1002/2014EF000252.

in sea surface height associated with the Gulf Stream's being close to the coast (also see discussion in Chap. 8). Variations in the strength of the Gulf Stream cause variations in the "tilt" of the ocean surface, and thus the height at the coast. A strong Gulf Stream tilts the surface away from the coast, and some estimates predict a slowing of the Gulf Stream. The slowing would result in enhanced sea level rise along the coast. There are systematic variations of sea level in Norfolk associated with internal modes of variability in the atmosphere, for example, the North Atlantic Oscillation. Sea-level rise sits in context with internal variability and other causes of relative sea-level change, which is typical of many applications.

Partly because the increase in flooding is widely obvious and interferes with commerce and day-to-day life, sea-level rise has received much attention in what is generally a politically conservative region. Residents, businesses, cities, and the military are all active in developing sea-level-rise policy and plans. Local universities have performed research quantifying the different causes of sea-level rise and contributing to communication of what has happened, framing vulnerability, and what is likely to happen. Many nongovernmental organizations (NGOs) with different focuses represent particular interests ranging from conservation to social justice.

In the mid-2000s, the need to address sea-level rise and climate change became so apparent that a regional focus started to emerge, with the local Intergovernmental Pilot Project being formalized in 2014.[9] This organization strives to coordinate the sea-level-rise preparedness and resilience planning of federal, state, and local government agencies and the private sector and take into account the perspectives of the region's citizens.

Though sea-level rise projections are often viewed as highly uncertain, the convergence of observations, people's perception of vulnerability, effective communication, and concern for the viability of the region stand as motivation to take action. With planning periods focused on the next few decades, building standards, zoning, and codes are being modified. There is recognition that at the end of this planning horizon, sea-level rise will not be stable, but likely to be increasing. Therefore, sustained, future-looking planning and design will be required. Climate model results provide a range of estimates of possible future states to inform this process (though they may be uncertain; see Chap. 11). However, climate change effects on sea-level rise are only one part of the planning process (see Chap. 12).

6.8 Summary

The ocean and ice portions of the climate system have vastly different scales: Sea ice is a tenuous and thin layer that exists between the atmosphere and ocean, but plays a huge threshold role in regulating climate at high latitudes, and through

---

[9]Center for Sea Level Rise, http://www.centerforsealevelrise.org/.

albedo feedbacks, affecting the planet globally. The ocean itself is a huge reservoir of heat that damps changes to the forcing applied to it. The ocean can be divided into a surface ocean above the thermocline (containing the mixed layer) and the deep ocean. In the deep ocean, formation of deep water is critical. For the mixed layer, surface fluxes and mixing are critical. Mixing occurs on smaller scales in the ocean than in the atmosphere, so ocean models are often run at finer scales, and mixing processes need to be well represented. Finally, the ocean supports a significant ecosystem (or set of ecosystems) that affect the carbon dioxide dissolved in the ocean and transferred either back to the atmosphere or into deep ocean sediments. The ocean is the link between the "fast" climate system (decades to centuries) and "geologic" timescales (millennia to millions of years).

Key Points

- The ocean is stratified, and density is important. Heat and salt control density.
- Ocean currents are driven by surface winds and the rotation of the earth, and deep currents are driven by density.
- Ocean models must represent small-scale eddies.
- Sea-ice models can represent recent losses in Arctic ice.
- The ocean carbon cycle is important for the global carbon cycle.

Chapter 7
Simulating Terrestrial Systems

We have discussed the atmosphere, the ocean, and the sea ice that floats on top of the ocean. The remaining major component of the earth system is probably the closest to you right now: the surface of the earth (unless you are reading this on a boat or a plane). The earth's surface is certainly closest to home. We can think of all the components of the climate on the (permanent) solid surface of the earth as the terrestrial system. Although this is commonly thought of as just modeling the land surface, it also includes two other important components: the cryosphere (ice and snow) that sits on land and the anthroposphere (the role of humans) in the climate system. We also discuss how human systems are simulated in general, and in climate models. Since all these interactions occur on the surface of the earth, the most useful way to discuss them is by looking at the land, cryosphere, and humans as parts of the *terrestrial system*.[1] Here we review the role of terrestrial systems in climate and discuss how they are simulated.

7.1 Role of the Land Surface in Climate

The surface of the earth plays several important roles in the climate system (Fig. 7.1).[2] Like the ocean surface, the land surface interacts with the atmosphere. The land surface also interacts with water and energy budgets. Land is only 30 % of the surface of the earth, but since humans are a terrestrial species, land has outsized importance to the climate system we experience.

---

[1]Lawrence, D., & Fischer, R. "The Community Land Model Philosophy: Model Development and Science Applications." iLEAPS and GEWEX newsletter, April 2013, http://www.cesm.ucar.edu/working_groups/Land/ileaps-CLM.pdf.

[2]For a review of many of these basic concepts, see Schimel, D. (2013). *Climate and Ecosystems*. Princeton, NJ: Princeton University Press; or Bonan, G. B. (2008). *Ecological Climatology: Concepts and Applications*. Cambridge, UK: Cambridge University Press.

© The Author(s) 2016
A. Gettelman and R.B. Rood, *Demystifying Climate Models*,
Earth Systems Data and Models 2, DOI 10.1007/978-3-662-48959-8_7

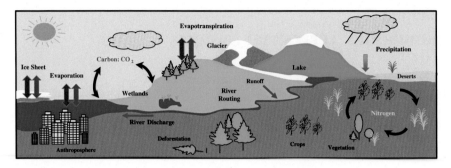

Fig. 7.1 Terrestrial systems and climate. Key processes include exchanges at the surface of water (evapotranspiration, precipitation, and evaporation) and energy (radiation). Key cycles of carbon and nitrogen are illustrated. Land surface processes of the hydrologic cycle are shown (precipitation, glaciers, runoff, and evaporation). Human disturbances (anthroposphere, deforestation, and crops) are also illustrated. Adapted from Lawrence and Fischer (2013)

7.1.1   Precipitation and the Water Cycle

Precipitation falls on the land surface. Some of this water is absorbed by plants and soils, some evaporates back into the atmosphere, and some may become surface water (streams, rivers, and lakes). The latter (runoff) occurs if the total water is greater than the holding capacity (or field capacity) of soil. The type of surface matters a great deal: Different land surfaces have different albedos, so they absorb radiation differently, and different soil types hold different amounts of water. As in the atmosphere, water in the soil is also important in the energy cycle. Heat may go into the land surface to evaporate water rather than heating the surface, and this has profound impacts. The fluxes (movements) of water are very important for recycling moisture back into the atmosphere.

7.1.2   Vegetation

Critical for the land surface is the role of vegetation. Vegetation alters the albedo of the land surface: Trees are darker than grasses, and grasses are usually darker than bare rock or dry soil. Vegetation changes the way that winds interact with the surface. But perhaps most critical are two aspects to vegetation that feed back into the climate system. First is the exchange of carbon, as plants "fix" carbon dioxide (CO_2) from the atmosphere via photosynthesis and incorporate it into organic matter. This fixation of carbon, driven by photosynthesis, is a critical part of the carbon cycle that ultimately helps regulate how much CO_2 is in the atmosphere. Second, plants use water in their tissues and for photosynthesis. In the leaves where photosynthesis occurs, some water leaks out, or "transpires" from the plants in a process called **transpiration** ("plant sweat," you might call it). Plants thus modulate the flux of water from the soil (where they take water up in their roots) into the canopy (the leafy region), where some of the moisture escapes into the atmosphere. Transpiration of water from plants is a significant fraction of the total evaporation: It is estimated to be a little over half of the total global **evapotranspiration** (the combination of water evaporated from the soil, water intercepted by leaves, and water released from plants via transpiration), and is a very important part of the water cycle. The role of animals in respiration (processing oxygen and CO_2) is too small to directly impact the carbon cycle. Think of the difference between the mass of animals in a square mile of natural land compared to all the plant material.

Changing vegetation can thus change the water available to the atmosphere. The change in atmospheric water can ultimately change precipitation downwind. We discuss the interaction between the land and atmosphere later in this chapter. Vegetation is not static, it will respond to changes in climate over time. The change in vegetation is often called **succession**, as one species or ecosystem gives way to another. The evolution of ecosystems creates a series of different feedbacks: As ecosystems change, the albedo and the ability to take up water and transpire water back to the atmosphere change. The different recycling of moisture can alter total precipitation. Thus, the vegetated land surface plays an important and active role in climate, particularly in the hydrologic cycle. But not all surfaces are vegetated.

7.1.3   Ice and Snow

Some of the most important land surfaces are covered with ice or seasonal snow, and/or contain permanently frozen soil (permafrost). These are the portions of the cryosphere that are on land, and they represent seasonal snow-covered tundra (with and without permafrost), mountain glaciers and the two large ice caps of Greenland and Antarctica. These frozen surfaces are important for a variety of reasons on a number of timescales. We have extensively discussed snow and ice albedo feedbacks, and the sensitivity of high-latitude climate to changes in the surface albedo. The ice sheets are stores of 70 % of the total freshwater on earth. The Greenland ice sheet stores the equivalent water to raise global sea level by 23 ft (7 m). Antarctica contains by itself 60 % of the earth's freshwater, which if it all melted would be equivalent to raising sea level by nearly 230 ft (70 m). These ice sheets are generally thought to be quite stable. The Antarctic ice sheet has been there in some form for millions of years.[3] But Greenland is melting rapidly at the surface and the margins now, and part of the Antarctic ice sheet (the West Antarctic ice sheet) actually rests on land below sea level, making it more susceptible to erosion, flow, and melting. Recent evidence has indicated that warmer water around the edges of the West Antarctic ice sheet (10 % of the total mass of Antarctica, or 23 ft, 7 m of

---

[3]Bender, M. L. (2013). *Paleoclimate*. Princeton, NJ: Princeton University Press.

sea-level rise) may be making it unstable.[4] Given the importance of these ice sheets, new modeling tools are being developed to simulate their dynamics.

7.1.4 Human Impacts

Finally, there is one other very important surface type: the regions of the earth's surface that are affected or regulated by humans. Likely you are in one of those regions now, and, by definition, you live in one of these regions. This includes regions of urbanization, and land use change caused by humans for agricultural use (crops and grazing) and wood harvesting (deforestation). Cropland covers about 10 % of global land area,[5] and pasture another 25 %. This is one-third of the earth's land area. Forests of all sorts are another 30 %. About a third of total land area is tundra, deserts, or mountains. So a significant fraction of the earth's surface is being used intensively by humanity.

Humans have altered about half of the **arable land** (land on which plants grow and animals can find food).[6] The land use by humans affects many of the cycles and feedbacks noted earlier. Shifting from natural vegetation to crops, with their need for additional water (irrigation) and nutrients (fertilizer, mostly nitrogen and some phosphorous) can alter nutrient cycling and can also drastically change the surface albedo as well as surface heat and moisture fluxes. For example, the rainforests in the Amazon, through evapotranspiration, recycle water back to the atmosphere that falls again as rain. Theories (and models) predict that as the rainforest is converted to pasture land (as is happening now), this might reduce the evapotranspiration and reduce rainfall, potentially making the remainder of the forest more vulnerable.[7] Thus, trying to understand the impacts of our actions on the land surface and how that might affect the climate system is important, whether global climate change or local land use change is a driver.

In addition to changes to the physical land surface, humans have significantly altered chemical and nutrient cycles (often called biogeochemistry; see Sect. 7.3)[8] through the activities of organized societies. Humans have perturbed the CO_2 concentration by about 40 % (from 280 to 400 parts per million) over the past

---

[4]Alley, R. B., et al. (2005). "Ice-Sheet and Sea-Level Changes." *Science*, *310*(5747): 456–460.

[5]Based on land-use data available from the World Bank, http://data.worldbank.org/indicator/AG. LND.ARBL.ZS, or the *CIA World Fact Book*, https://www.cia.gov/Library/publications/the-world-factbook.

[6]For arable land trends over time, see United Nations Food and Agriculture Organization Statistics division (FAOSTAT), http://faostat3.fao.org/home/E.

[7]Malhi, Y., et al. (2009). "Exploring the Likelihood and Mechanism of a Climate-Change-Induced Dieback of the Amazon Rainforest." *Proceedings of the National Academy of Sciences, 106*(49): 20610–20615.

[8]For a detailed background, see Charlson, R. J., Orians, G. H., & Butcher, S. S. (1992). *Global Biogeochemical Cycles*, ed. G. V. Wolfe. New York: Academic Press.

several centuries. But we have also perturbed the cycle of fixed nitrogen (an important nutrient and the primary nutrient in fertilizer) by at least 100 %. The particles we emit from industry or as by-products of energy use may affect how clouds form and how much energy is absorbed in the atmosphere, and chemical perturbations (air pollution or smog) damage animals and plants. So our economic systems are coupled to the climate system.

There are models that try to simulate these human systems and their two-way interactions with climate. For example, as temperature rises, humans use more energy to cool their societies (think of Arizona or Athens on a hot day). This increased energy use will have economic costs and environmental impacts (more energy, more CO_2 emissions). More CO_2 emissions will further impact climate. So coupling models of economic systems to climate is also critical, and the land surface is where that happens.

7.2 Building a Land Surface Simulation

The terrestrial surface is simulated by representing a series of these critical processes, on the surface of the earth. The discussion also tracks the history of development of land surface models. First comes the treatment of exchanges with the atmosphere and ocean: surface fluxes and heat, and also a discussion of **hydrology** (the water cycle). This includes the water in the soil, and how it evolves. Together, these heat and moisture fluxes are often called **biogeophysics** (*bio* = living things, *geo* = earth). The critical *bio*-part of biogeophysics includes how vegetation alters fluxes of heat and moisture through evapotranspiration. Surface waters (rivers, lakes, and wetlands) are also important for hydrology.

The next phase of terrestrial system modeling involves nutrient cycles, chiefly of carbon and nitrogen, but also some minor species important for plant growth. In addition to these nutrient cycles comes a representation of the dynamic land surface: changing land cover types including cropland, simulating disturbances (deforestation or fires), and even models of the urban system.

The cryosphere is also important in terrestrial systems. Snow is an important surface type. Recently climate models have started to develop new models for "land ice" (glaciers and ice sheets). Finally, we mention models of human social and economic systems.

7.2.1 Evolution of a Terrestrial System Model

Terrestrial system models generally began as a set of parameterizations to consistently supply energy and moisture fluxes to the atmosphere. A basic land surface model contains timescales and efficiencies (often characterized as resistances to flow, as in an electric circuit) for water and energy. Basic models also include simple representations of how much water the soil can hold, and its exchange into the atmosphere.

The next step in complexity is to treat soil and plant systems in more detail, with representations of different soil and plant types. This is important for representing the effect of plants through evapotranspiration. Soil and plant submodels have been extended so that the vegetation can evolve over time in response to nutrients (bio-geochemistry) and climate, so that the vegetation becomes dynamic. What is dynamic vegetation? It means the type of vegetation is not fixed and can evolve as the climate changes. For example, if precipitation changes, rainforests can turn into grassland.

The last stages of complexity are to also simulate the effects of humans, and a complete land cryosphere. Human perturbation to climate can be integrated into the system with economic models coupled to the physical models. Attempts to represent more completely glaciers and ice sheets are also under way.

It is clear that a terrestrial system model is really a system of coupled submodels (land ice, human systems, vegetation types, soil, hydrologic cycle) that represent key processes that occur on the land system.

One important requirement of terrestrial system models is the ability to represent the different impacts of small-scale features—such as forests versus grasslands versus lakes—within a large-grid cell. One advantage of terrestrial system models is that generally soil and plants do not move in the horizontal. Only water really moves rapidly in the terrestrial system. This improves the computational efficiency of terrestrial models. It also helps that land represents only 30 % of the area of the earth.

Land models have a different approach to subgrid variability than do models of the ocean, atmosphere, or sea ice. While things like clouds in part of a grid cell are ephemeral and change rapidly, the proportion of a part of the earth's surface (say, 62 miles or 100 km on a side) covered with a given type of soil and/or vegetation does not change much. So land models often split up each grid cell into the different vegetation and surface types (lakes, urban areas, and glaciers) that might be present at any location. Fluxes of water, energy, and carbon are then calculated separately for each surface type and then the grid-cell fluxes are calculated via a weighted average based on the proportion of each surface type within the grid. The distribution of surface types remains relatively fixed, evolving only slowly over time due to human alterations of the land surface or through vegetation disturbances due to fire or a response to a climate shift. Unlike the atmosphere, there is very little communication between different grid points or even subregions of the grid at the surface. Generally there is only runoff of surface water. Conduction of heat and subsurface flow is not treated in most large-scale terrestrial system models. Mostly individual land units exist independent of the others, as if the model were an atmospheric general circulation model (GCM) with no dynamics (i.e., no general circulation) and advection, and just a series of columns with a list of different surface types in each column.

Topography, defined as the change in elevation, generates additional complexity in terrestrial system modeling. When subgrid variations of the surface are considered for different surface types (soil or vegetation) on scales of hundreds of kilometers, there are also elevation changes to consider. This requires addressing variations in temperature and precipitation, and the treatment of the angle and slope of terrain. As an example, a grid box with a mountain range will have very different land cover types as altitude increases from forest to alpine to snow/ice or bare rock. Also, the windward side of the range (facing the prevailing wind) will generally have more precipitation and different vegetation types than the leeward (downwind) side, which often sits in a "rain shadow." Fortunately, many of these topographic features are fixed, so relationships for these distributions can be designed (parameterized). For example, the mean surface temperature of a grid box can be distributed so that some regions have warmer and some colder temperatures depending on elevation. Precipitation can be distributed unevenly based on the direction of the wind. These complexities are necessary for getting local climates and land surface types correct.

7.2.2 Biogeophysics: Surface Fluxes and Heat

Atmosphere models have a bottom boundary, and the first terrestrial system models were really just surface flux parameterizations that represented the heat exchanged at the surface. Radiation impinges on the surface both as direct solar radiation and

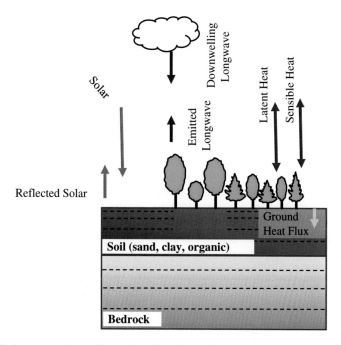

Fig. 7.2 Surface energy fluxes. Key surface fluxes into and out of the land surface. Radiation from the sun (solar) is in *orange*. Infrared (longwave or terrestrial) radiation is *red*. Fluxes of water (latent heat, precipitation) are *blue*. Sensible heat exchange is *magenta* and the heat into the soil is *yellow*. Adapted from Lawrence and Fischer (2013)

Fig. 7.3 Bucket model. Soil can hold water based on precipitation input and evaporation loss up to its field capacity. Water in excess of the field capacity becomes runoff

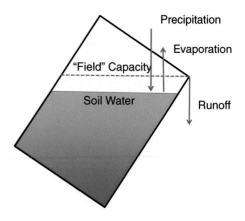

as more diffuse (backscattered) solar and longwave (terrestrial) radiation. Some of this energy is not just in the form of heat, but also energy that comes from the potential to condense water (called **latent heat**), which ties the **energy budget** to the hydrologic cycle. Figure 7.2 illustrates the absorption and reflection of energy, and the latent and sensible heat. These surface fluxes then send heat into (and out of) the ground and soil as the ground heat flux.

7.2.3   Biogeophysics: Hydrology

In addition to surface energy fluxes, the flow of water in and out of the soil is represented. One common form of hydrology, originally conceptualized in the late 1960s in the Soviet Union, is to treat the soil column as a "bucket" that can hold a given amount of water. The precipitation, minus the evaporation, is used to fill or empty the bucket, and if the bucket of soil gets to a threshold where it cannot hold any more water (the soil's **field capacity**), the water runs off. Figure 7.3 illustrates a **bucket model**.[9]

Soil moisture is a critical part of the climate system. It regulates what happens to precipitation, and how it gets recycled into the atmosphere, stays in the soil where it is available for plants to use, or becomes surface runoff. There are important feedbacks between the atmosphere and the soil moisture as well.[10] Wetter conditions (more precipitation) mean more evaporation, and a wetter atmosphere above the land. Evaporation is latent heat: Energy goes into evaporating water rather than increasing temperature, and this latent energy dominates in wet conditions. With

---

[9]The original treatment of the Budyko bucket model is reviewed in Budyko, M. I. (1974). *Climate and Life*. New York: Academic Press.

[10]For a review of soil moisture feedbacks, see Seneviratne, S. I., et al. (2010). "Investigating Soil Moisture—Climate Interactions in a Changing Climate: A Review." *Earth-Science Reviews*, 99(3–4): 125–161.

Fig. 7.4 Hydrology of the terrestrial model. Precipitation comes from the atmosphere. Evaporation, sublimation (evaporation directly from snow), and transpiration (from plants) returns water to the atmosphere. Water also goes into soil and surface runoff. The soil water can also recharge an aquifer in the deep soil layer. Adapted from Lawrence and Fisher (2013)

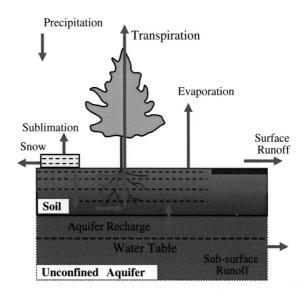

wetter conditions, the surface is energy limited (more energy is needed to evaporate water). The only way that water can evaporate is by adding more energy to the system. In wet conditions, adding more energy does not necessarily increase the temperature; rather it can just go into latent heat, evaporating more water.

Drier conditions are often water limited, meaning there isn't enough water to satisfy the energy demands. In dry regions, there is basically very little vegetation and little transpiration. In between, in semi-arid regions, the transpiration (evaporation from plants) is important. Transpiration can be dependent on soil moisture. Less soil moisture means more sensible heat and larger increases in temperature for a given energy input. Thus, the surface properties and biogeophysics of soil moisture and total evapotranspiration, even diagnosed with a simple model, show the importance of the land surface for local climate.

The simple bucket approach is quite useful, but it has many limitations. It is basically the simplest approach to capture some of these atmospheric feedbacks with the surface. More modern treatments of the soil include multiple layers, and water transport across those layers, in the subsurface soil. Many of these processes are illustrated in Fig. 7.4. Having multiple layers and different pathways through the soil for local hydrology allows for a better representation of the variability in soil environments: from permafrost (with variable frozen layers), through to tropical wetlands with saturated soils. The **tiled approach** allows many soils to be present in a single grid box with varying fractions and effects, for example, like only part of the grid box's being saturated.

When the soil becomes saturated with more water than can be absorbed, runoff occurs, generating lakes, rivers, and wetlands. The bucket model is a simple illustration of this. Surface water has often not been treated extensively in terrestrial system models. But of course lakes and rivers are important for climate. Rivers move freshwater into the oceans. They also move nutrients into the oceans (see Sect. 7.2.4). Lakes and wetlands, with potential large area coverage, are also important for providing regions of large-scale evaporation, with significant effects on climate: Surface water can readily evaporate back into the atmosphere. Regionally this can be a dominant source of water for precipitation. This is familiar to anyone who lives to the east (downwind) of the Great Lakes region of North America (e.g., western Michigan, parts of New York state), where "lake effect" snow storms result from moisture picked up as cold air flows over large lakes that are relatively warm. Lake water evaporates into the air and then this water is deposited downstream as snow when the air cools. Representing lakes in climate models is critical for getting regional climate correct. Human modifications of surface water systems, via dams and reservoirs, also create such lake effects from evaporation, and may modify local or regional climate. Though human-made lakes are usually much smaller than the Great Lakes, the effects can still be important locally. Plus, these reservoirs store and evaporate water, thereby altering river flow.

The deepest piece of the land hydrology is the storage of water in aquifers beneath the soil. **Aquifers** are regions of permeable rock containing groundwater that can be extracted with a well. A common analogy is digging a hole in the sand at a beach: When water is reached, it flows into the hole (a well). The level of the water is the **water table** and the moist sand the aquifer. Geology creates these regions with permeable soils, and they are "recharged" by seepage of groundwater into them from precipitation. Representing "stored" water, aquifers are an important part of the hydrologic cycle: They can be used to provide water when no surface water or precipitation is available. Most regions of the earth have some sort of aquifer beneath them. Aquifers are critical for human populations. Many human settlements coalesced around wells, and a lot of agricultural areas are dependent on groundwater for irrigation. Aquifers typically interact with the soil in land models, forming a deep storage region for water that penetrates the soil. Simulating aquifers is important for understanding the low-frequency behavior of hydrology and the interaction of hydrology with humans on climate scales.

7.2.4 Ecosystem Dynamics (Vegetation and Land Cover/Use Change)

Representing vegetation is an important part of modern terrestrial system models. Proper representation of vegetation is important for biogeophysics: heat and moisture fluxes between the atmosphere and the surface. As we have seen, this is particularly because of transpiration from plants. The structure of a vegetation canopy also is important for regulating how surface radiation fluxes filter between the top of the canopy and the surface. Vegetation creates its own near-surface boundary layer that modifies surface fluxes. And vegetation also is important for nutrient cycles, including the cycling of carbon between organic matter in the land surface and the atmosphere (see Sect. 7.3).

Vegetation also has subgrid-scale variation. Typically a model will have different vegetation classes, or **plant functional types**, in a single grid cell. These are probably best thought of as simple representations of different ecosystems. For example, a grid cell might contain forest, grassland, cropland, and/or tundra. The characteristics might be more detailed: There may be several different types of forest (e.g., broad-leaf deciduous forest, evergreen forest). Different plant types behave differently with respect to the properties of transpiration, leaf area, canopy height, and so on, so it makes sense to have different "tiles" or units that can represent the different ecosystems. Modern terrestrial system models can have 10–15 of these plant functional types. The types classify an ecosystem of plants by plant traits both above ground (height, flammability, leaf area, and nutrient content) and below ground (root depth, nutrient uptake). Plants are also classified by their different physical traits: their ability to grow and use nutrients and water (called plant **phenology**).

A great deal of detail in current terrestrial system models is being added at the fundamental level of understanding plant physical traits. This is the description of the physical nature of plant characteristics for things like transpiration, growth (uptake of carbon), even their albedo. Properly describing these traits and then having multiple plant types enable a complex treatment of the land surface with quite a bit of diversity. These plant characteristics are typically based on direct observations of plant types. The goal of the description of the plant functional types is to represent how the environment affects plant growth, and how plants in turn affect the environment.

When vegetation and vegetation characteristics were first introduced into surface models, the description of the vegetation was taken to be static: not unlike the topographic conditions describing the arrangement of the land surface and its elevation. Similarly, the first atmospheric models (in the 1960s and 1970s) had fixed cloud distributions (see Chap. 5). Even a static distribution of vegetation, with different properties for on the order of 10 different ecosystems, is a significant improvement compared to no vegetation. It allows for a more complete representation of important properties of the surface system (like evapotranspiration). But vegetation is not static, and vegetation health and distribution can evolve in response to disturbances (natural or human caused) and due to climate change. Disturbances include occurrences such as fire, disease or insect outbreaks, and drought. Climate changes can make the present distribution of vegetation unable to survive, and cause succession of one plant type into something else. The representation of dynamic vegetation in terrestrial systems is often called **ecosystem dynamics**. In this context, dynamic means change in a temporal sense: changes over time.

Terrestrial system models now commonly include representations of how vegetation distributions could evolve. This is illustrated in Fig. 7.5. Some changes are natural, or a response to climate: Plants will die off or grow better depending on climate changes. If the planet warms or cools, certain vegetation classes will

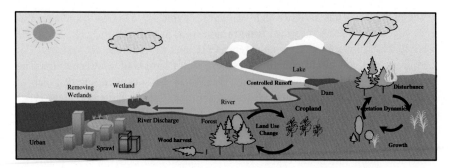

Fig. 7.5 Ecosystem dynamics. Ecosystems can be disturbed (e.g., from fire), and they evolve and change ecosystem type. Ecosystems can have humans change their type (e.g., to cropland or through deforestation). Runoff and rivers can be controlled, wetlands can be altered, and urban landscapes expand. Adapted from Lawrence and Fisher (2013)

populate different regions. This can also occur due to disturbance. **Natural disturbances** can be fires or floods. **Human disturbances** are often called "land use change," such as forest conversion to crop or pastureland, or even conversion to urban environments. For natural changes (responses to climate change, for example), rules can be developed for the success or failure of each plant functional type for different climate regimes, and the evolution (or succession) of those types. One example might be the disturbance of a forest by fire, and the conversion of the forest to grassland, then to a deciduous forest and then to an evergreen forest over time. While we generally think of human land use change as something we must impose, there are ways to try to simulate the evolution of human systems (Sect. 7.6) and how they might alter land use (Sect. 7.7). Thus, ecosystem dynamics is also the link to human systems and their impact on the land surface.

7.2.5 Summary: Structure of a Land Model

The structure of a land model grid cell describes how the model interacts with the atmosphere above it, through exchanges of heat and moisture (see Sect. 7.2.2). Generally, the land surface will be broken up into different surface types, or ecosystem types, such as desert, grassland, or forest. Multiple types can exist in the same grid cell. The model will have a description of the characteristics for the soil properties in each grid cell, and possibly for each surface type. This includes a description of the hydrology of the soil: how much water it can hold (the field capacity; see Sect. 7.2.3). It also accounts for excess water that may run off.

Each surface or ecosystem type will have a description of the plants on the surface (see Sect. 7.2.4). The description includes a description of how an ecosystem of plants (like the trees in a forest) moves water into the atmosphere through evaporation, reacts to precipitation, and grows and decays. This helps to determine the surface exchanges of heat and moisture and also allows a description of the carbon content of the soil and plants.

All of these characteristics are input to the model. The plant responses to the environment are usually derived from observations of actual plants: either single plants or detailed measurements of entire ecosystems such as forests or grasslands. The data are distilled down into relationships. If the rainfall is W and there is at least X amount of nitrogen and water in the soil, then the plants will grow in that ecosystem and they will take Y amount of carbon from the atmosphere and Z amount of carbon from the soil. This is calculated for every ecosystem type in a grid cell, and for every grid cell in the land model. The moisture in the soil is estimated from a hydrology model. Excess surface water becomes runoff that flows into the next (downhill) grid box.

Now we focus on some of the key nutrients that limit plant growth, and on the flows of carbon that go into plants. Carbon is of concern because it exchanges between the soil, plants, and the atmosphere. In the atmosphere, carbon is CO_2, the greenhouse gas.

7.3   Biogeochemistry: Carbon and Other Nutrient Cycles

Plants affect the water and energy fluxes at the surface through transpiration, and through canopy absorption and emission of radiation. These are immediate effects that affect weather as well as climate scales. Plants are also important for cycling nutrients: key chemicals in the earth system on which life depends. Nutrients cycle through the earth system, and understanding the flow of these nutrients is called **biogeochemistry** (a complement to biogeophysics).

The role of **nutrient cycling** is important for two reasons: Some of these chemicals directly affect climate, and others affect plant growth in the ocean and on land. The critical nutrients we focus on are carbon and nitrogen. Carbon is an example of a chemical that affects climate as carbon dioxide (CO_2) or methane (CH_4) in the atmosphere. Nitrogen is an example of a critical chemical affecting plant growth. Phosphorous is also an important nutrient for plant growth, especially in tropical ecosystems.

Biogeochemical cycles describe what happens to key nutrients in the earth system. The key concept is a cycle: There are a series of **reservoirs** in which carbon exists in different forms. The **carbon cycle** is illustrated in Fig. 7.6.[11] Carbon reservoirs include rocks and minerals (including geological storage of carbon in fossil fuels), the ocean (where carbonate minerals and CO_2 are dissolved in the water column), vegetation and soil, and the atmosphere. The figure also indicates the exchange between the reservoirs, and the changes to the exchange between

---

[11]An accessible introduction to the carbon cycle is Archer, D. (2010). *The Global Carbon Cycle*. Princeton, NJ: Princeton University Press.

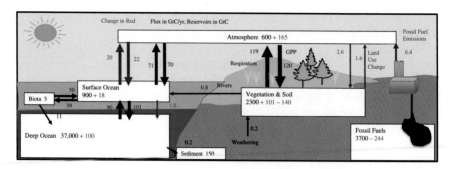

Fig. 7.6 The carbon cycle. The largest climate system reservoirs for carbon include the deep ocean, soil and vegetation, surface ocean, and atmosphere. The approximate size of annual carbon fluxes is given by the width of the *arrows*; *red arrows* indicate perturbations by humans. *Black arrows* are natural exchanges. Quantities are in gigatons (10^9 tons) of carbon (GtC) and gigatons of carbon per year (GtC/yr)

reservoirs induced by human activity. With all biogeochemical cycles, it is important to understand not just the size of the reservoirs, but also the fluxes between them and the "lifetime" of the reservoir turnover. That is why a "small" increase in CO_2 from fossil fuels (1 % per year increase) can build up in the atmosphere. Similar cycles can be drawn for a number of different important nutrient species.[12] Earth system models are now starting to represent these reservoirs and fluxes in the various components.

The terrestrial biosphere (soil and vegetation) is the largest gross exchange of carbon (often called the **gross primary productivity**, or GPP) with the atmosphere. The sink of carbon from the atmosphere to the land cannot be measured but is typically calculated as a residual. Nearly half of the CO_2 we emit from fossil fuels and land use change (deforestation) stays in the atmosphere. Observations indicate that a bit less than one-quarter of the additional CO_2 is going into the ocean.[13] This leaves about one-quarter of human emissions to go into the land surface. Thus, half the additional CO_2 is flowing through the carbon cycle and leaving the atmosphere (where it does not function as a greenhouse gas to warm the planet). One of the big outstanding scientific questions in the field of biogeochemistry is whether this partitioning of the **carbon sink** from the atmosphere will continue. If forests die, or if the carbon trapped in

[12]An overview of other trace element cycles is found in Jacobson, M., Charlson, R. J., Rodhe, H., & Orians, G. H. (2000). *Earth System Science: From Biogeochemical Cycles to Global Changes*, Vol. 72. New York: Academic Press.

[13]Takahashi, T., et al. (2002). "Global Sea—Air CO_2 Flux Based on Climatological Surface Ocean pCO_2, and Seasonal Biological and Temperature Effects." *Deep Sea Research Part II: Topical Studies in Oceanography, The Southern Ocean I: Climatic Changes in the Cycle of Carbon in the Southern Ocean*, 49(9–10): 1601–1622. doi:10.1016/S0967-0645(02)00003-6. For a classic overview of the carbon cycle and sinks, see Siegenthaler, U., & Sarmiento, J. L. (1993). "Atmospheric Carbon Dioxide and the Ocean." *Nature*, 365(6442): 119–125.

permafrost is released, the carbon budget may change. One of the key goals of land models is to simulate the carbon cycle and predict whether and how it might change.

The land surface is an important part of the carbon cycle that helps determine atmospheric CO_2 levels (and hence climate). Organic matter contains a great deal of carbon. Organic molecules generally contain a ratio of carbon to hydrogen to oxygen (C:H:O) of 1:2:1. Understanding and representing the biogeophysics of plants and soil is the first step to determine the temperature, moisture, and plant types, from which the carbon contained in plants can be estimated, and its evolution through soil and exchanges with the atmosphere can be modeled. The reason why plants transpire water is because their stomata (like pores, but in leaves) open to allow the exchange of CO_2 and oxygen for photosynthesis, and some water is lost. Increasing concentrations of CO_2 in the atmosphere can reduce the need for plants to open their stomata and lose water, because it is easier for the plant to collect CO_2 from the atmosphere. The change in plant physiology with higher CO_2 can change the growth rate of plants, and their primary productivity, thus changing the sink of carbon.

The response of plants to increasing CO_2 is often called **CO_2 fertilization** or the **carbon cycle feedback**. One hypothesis regarding increased land uptake of carbon is that plants are more efficient at growing with higher CO_2 concentrations and can grow more, pulling more carbon into their tissues. But increasing temperatures and changing precipitation are confounding factors that may limit the efficiency of plant growth. Thus,

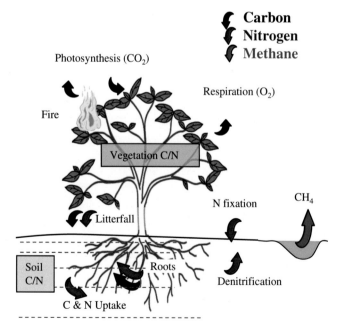

Fig. 7.7 Biogeochemical cycles in a terrestrial system model. Vegetation carbon and nitrogen go into the soil or the atmosphere. Carbon is produced in plants from photosynthesis. It leaves plants when they burn or decay, the latter carbon going mostly into the soil. Nitrogen is fixed and removed (denitrification) from the soil. Methane is produced in wet (anoxic: no oxygen) environments. Adapted from Lawrence and Fisher (2013)

understanding the carbon cycle, and representing it, becomes critical for the evolution of atmospheric CO_2 and the climate system. The cycle is built up from careful measurements of how plants grow and process nutrients (physiology), how they compete and evolve, and then how they respond to climate and climate changes. Chapter 11 discusses some of the predictions of the carbon cycle from current climate models.

Terrestrial models are now including representations of the transport of carbon in the system with representations of the carbon cycling, sometimes called biogeochemistry. An example is shown in Fig. 7.7. Vegetation and soil carbon are primary reservoirs. Key processes for carbon are photosynthesis, respiration by plants, and decay in the soil. Methane (CH_4) is important in aquatic or inundated ecosystems, where methane is produced by bacteria in the absence of oxygen.

But carbon is not the only biogeochemical cycle that is important for climate. Organic matter contains and requires other elements as well. After carbon, organic matter contains nitrogen, in a ratio varying from C:N of 106:16 in the ocean to 160:1 for plants, and 15:1 for soil organic matter. Nitrogen gas is the largest component of air, but it is inert and can be converted for organic use ("fixed") by only a few plants and microbes in soil, symbiotically living with plants, or some algae. Specifically, atmospheric nitrogen (N_2) is converted into ammonia (NH_3) by an enzyme. The absence of this fixed nitrogen can limit plant growth. Hence, representing nitrogen is important for understanding how plants will use water and carbon. Figure 7.7 also shows some important terrestrial processes for the **nitrogen cycle**,[14] including fixation (making nitrogen into forms used by plants) and uptake in soil, and leaching and loss of fixed nitrogen in soils. There are a whole host of other nutrients that play a role in the growth of ecosystems in small concentrations. Phosphorous and iron are the next most important elements, and iron may be a limiting nutrient in some ocean ecosystems, as it can arrive only by dust deposition to the ocean surface.

The Carbon Cycle

Carbon is magic stuff. It makes up organic matter, whether living plants (including algae in the oceans), organic matter from dead plants (or algae) in the soils or on the sea floor, or animal tissue (our bodies, or plankton in the ocean). The carbon that is in our bodies or in plants comes ultimately from minerals in the earth, but it often arrives by being a gaseous species in the atmosphere. The carbon dioxide and methane that are greenhouse gases regulate and alter the radiative energy leaving and entering the earth system. Carbon dioxide is used by plants in photosynthesis and is a by-product of the respiration process used by animals for energy, returning to the atmosphere or mineral form when we are done using it (and plants or animal tissues decay). This cycling is critical for understanding how the climate evolves on long timescales. And by long, that can mean geological timescales (up to millions of years).

---

[14]Galloway, James N., et al. (2008). "Transformation of the Nitrogen Cycle: Recent Trends, Questions, and Potential Solutions." *Science, 320*(5878): 889–892.

The carbon cycle is one reason that climate change is a difficult problem to understand, predict, and simulate. For many environmental issues, a human-induced chemical, compound, or process is introduced into the earth system that did not exist before, for example, the refrigerants called **chlorofluorocarbons** (CFCs) that cause depletion of stratospheric ozone, or chemicals like Dichloro-diphenyl-trichloroethane (DDT) that cause cancer in humans and animals. These chemicals do not even sound natural. But with climate change, the "culprits" are a part of the system itself: Carbon dioxide is literally the breath of life for plants, and it is part of our bodies and the food we eat. Like many things, it is the change in natural cycles, "too much of a good thing," that potentially alters the system. And because the systems are delicately balanced and coupled (carbon dioxide affects transpiration, which alters water fluxes at the surface; e.g., see Fig. 7.6), large climate changes can result.

The carbon cycle is in a delicate balance, and is coupled to the climate system on many timescales. Because the two most important greenhouse gases after water vapor are part of the carbon cycle (carbon dioxide and methane), carbon has a very direct effect on the radiation absorbed and emitted by the earth, and hence the global climate.

The different reservoirs in the carbon cycle (see Fig. 7.6) have different timescales: from days to years in the case of plants and soil, to hundreds of years for the atmosphere and deep soil, to thousands of years for the ocean, and millions of years for weathering and sediments. The feedbacks in the system are complex. Carbon moves in and out of the atmosphere, and into other reservoirs. Slow feedbacks from weathering and burying carbon in sediments are different than many of the faster feedbacks between land and atmosphere. The ocean and land are currently thought to be taking up more carbon than they release because of the increase in atmospheric carbon dioxide. The land can change much faster than the ocean, because it is more prone to disturbance, and the ocean acts as a big damper on the system. A key current goal of earth system models is to represent flows and reservoirs of carbon in the earth system. In the atmosphere, it is fairly simple to represent carbon containing gases and particles. The land surface and ocean require extensive treatment of their biology to cycle carbon through the systems and determine its fate. These biological cycles and models are some of the most important and uncertain parts of earth system models, and representing their feedbacks becomes important on long (century or more) timescales.

7.4 Land-Atmosphere Interactions

From the descriptions in Sect. 7.3, you can see that there are several important ways in which the land surface is coupled to the atmosphere and can affect the atmosphere directly. The effect can certainly be on climate scales, such as the scale of the carbon cycle and the CO_2 fertilization effect. But the coupling can be short term and more process based, such as the land surface cycling and processing of water through the surface hydrologic cycle. Precipitation falls on the land surface, and can take several pathways. Surface water can evaporate back into the atmosphere. This changes the climate in some regions by providing a water source. But subsurface water in the soil (soil moisture) can be recycled back into the atmosphere by transpiration from plants. This can directly affect the low-level atmospheric humidity, and recycle humidity, for example, back into the lower atmosphere, where it can alter cloud formation. The humidity and soil moisture can also alter the partitioning of surface fluxes between sensible heat (increasing temperature) and latent heat (increasing evaporation).

Altering the local hydrologic cycle can affect short-term weather systems, or long-term climate. One storm may moisten part of a land surface. The runoff or soil moisture provides future evaporation back to the atmosphere, but it also can reduce temperatures when that evaporation occurs. On the climate scale, the moisture recycling of terrestrial systems may maintain certain climate zones. The Amazon rainforest is often called the **green ocean**. This is partially because rainforest plants provide a large additional source of humidity back into the atmosphere that can rapidly recycle rain back into the system. Many models do not capture this effect fully. If the Amazon exists because of these couplings between land and atmosphere, it may not be stable to large-scale disruption. Many coupled climate models have trouble maintaining the Amazon rainforest: It tends to dry out and turn to grassland if precipitation is a little too low. This highlights the importance of understanding and representing land-atmosphere coupling in global climate models.

7.5   Land Ice

Snow- and ice-covered surfaces are other large and long-lived parts of the terrestrial system with important implications for climate.[15] Ice sheets include Antarctica and Greenland but also some small ones in Iceland and other regions. There are also numerous glaciers in high latitudes and high altitudes distributed around the planet. But it is really Greenland and Antarctica that contain large amounts of ice. The cryosphere on land, like that contained in sea ice, is important for its effect on the earth's albedo. The land-based frozen water is also important for the storage of freshwater and regulation of sea level. In addition to the rapid decrease in Arctic sea ice and seasonal snow cover, there is increasing evidence of surface melting and increased glacier flow in Greenland. The increased awareness of the potential risk of significant melt, or catastrophic collapse (which would raise sea level, and change the ocean density in a region of deep water formation), has motivated detailed observations and modeling studies. The simulation of ice sheets on land is

---

[15]An overview of the role of the cryosphere in climate is contained in Marshall, S. J. (2012). *The Cryosphere*. Princeton, NJ: Princeton University Press.

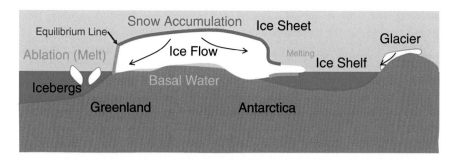

Fig. 7.8 Glaciers and ice sheets. Greenland (*left side*) has ablation (melt) on the lower parts and accumulation on the upper parts. Basal (base) water lubricates ice flow. Antarctica (*right side*) has ice shelves with melting beneath, and some of the ice sheet is grounded below sea level. Glaciers (*far right*) have similar processes including ice flow

still evolving rapidly, and many climate models do not fully treat ice sheets. Glaciers are typically not treated at all. In the absence of an ice sheet model, the ice-covered regions are just treated as (permanently) snow-covered land.

Terrestrial systems often have fairly sophisticated models of snow that falls on land, ice, and sea ice. This is because snow serves as an intermediate layer that mediates surface energy and water fluxes between the atmosphere and land or ice surface. Snow has a different albedo from land or ice, and the albedo can change over time as the snow crystals "age." Snow models generally have several layers and a complex representation of radiative fluxes. Snow models also treat deposition of particles on their surface: Particularly important in some regions is soot (black carbon) particles from fires and industrial activity, or mineral dust from far-away deserts. These dark particles can significantly lower snow albedo, resulting in more absorption of solar energy. This means more snow melt, and it can accelerate albedo feedbacks at high latitudes. Snow also acts as a strong insulator, keeping high-latitude soils much warmer through the winter when snow is present than when it isn't. In fact, changes in snow depth due to climate change can have as big an impact on soil temperatures as climate warming itself (either amplifying or offsetting climate change, depending on whether snow depths increase or decrease).

Glaciers and ice sheets have many of the same surface properties, and indeed, a snow model would commonly also run where there is an ice sheet or glacier (annually retained ice at the surface). But ice sheets and glaciers have other components of surface melting and accumulation in addition to radiative fluxes at the surface. Ice sheets can also be present for a long and stable period of time. Ice cores dating back 800,000 years and over 2 miles (3 km) deep have been retrieved from Antarctica.[16] Ice sheets and glaciers also move slowly. Simulating ice sheets requires simulating the accumulation and melting process as well as the ice flow.

---

[16]There is an 800,000-year record from the "Dome C" ice core. Original results are reported in Lüthi, D., et al. (2008). "High-Resolution Carbon Dioxide Concentration Record 650,000–800,000 Years Before Present." *Nature, 453*(7193): 379–382.

Simulating the flow is critical and difficult, because ice sheet and glacier flow depends strongly on the melting and the pressure on the bottom (bed) of the glacier.

Antarctica and Greenland have slightly different processes, illustrated in Fig. 7.8. Greenland has a large region at lower altitude where the ice is melting (ablation) for part of the year. Greenland also has ice streams that run into the sea and calve (creating icebergs) at isolated points around the edge. There is also significant melting that flows down into the ice sheet forming "basal water", potentially lubricating the base for faster flow. The overall balance of an ice sheet depends on the balance between the accumulation (snow) on one hand and flow and melt on the other. There is an "equilibrium line" where the accumulation balances melt. Usually, melt dominates at lower elevations, and accumulation dominates at higher elevations (see Fig. 7.8). For Greenland, this line appears to be rising (more melting over more of the ice sheet).[17]

Antarctica is different. Temperatures are cold enough so that accumulation occurs across the entire ice sheet. In fact, warmer temperatures due to increased greenhouse gases will not bring the temperature above freezing. But they will allow more water vapor in the air, which will still fall as snow, and potentially lead to more net accumulation over the ice sheet. Antarctica has large ice shelves and even portions that are grounded below sea level, held in place by the weight of ice above. Ice sheets are supplied by flow from the interior, and they lose mass by iceberg calving at their face, but also by subglacial melting. If warmer water occurs underneath the ice shelf, this can erode the shelf in a sudden collapse, increase the flow, or even make the grounding line retreat (so that more of the ice sheet is floating and less stable). Recent analysis of the West Antarctic ice sheet indicates that such melting may already be happening. Recall that the West Antarctic ice sheet accounts for about 10 % of the Antarctic total ice mass, or about 23 ft (7 m) of sea-level rise. The only good news is that this may take several centuries to happen.[18]

The equations to model all of these ice sheet processes are relatively straight-forward applications of **deformable solid mechanics**, the description of how a solid material acts when force is applied or the temperature changes. They must also represent the different processes that occur, including subglacial water and melting from the bottom of ice shelves. Land ice models typically are simulated with long time steps (a season or a year) so that they can be run for thousands of years.

Including these ice sheet models in global climate models has begun recently and is an ongoing task, made urgent by the continued buildup of greenhouse gases and warming high latitude (especially Arctic) temperatures, combined with observations of significant increases in melting for Greenland. This all adds up to potentially large changes in sea level over the 21st century, hence, the desire to include ice sheets in climate models as a key part of the terrestrial system.

[17]Van den Broeke, M., et al. (2009). "Partitioning Recent Greenland Mass Loss." *Science, 326* (5955): 984–986.

[18]For a summary of the West Antarctic ice sheet, see Oppenheimer, M. (1998). "Global Warming and the Stability of the West Antarctic Ice Sheet." *Nature, 393*(6683): 325–332.

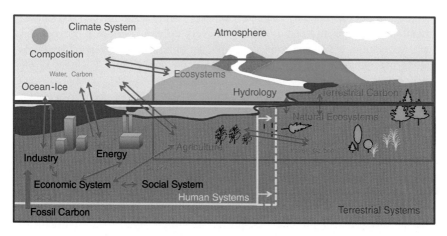

Fig. 7.9 Humans and integrated assessment. Interactions between human and natural systems. Human systems are outlined in *yellow*; interactions with terrestrial systems, in *purple*; natural ecosystems, in *green*; and the climate system, in *blue*. Fluxes of water (*blue arrows*) and carbon (*orange arrows*) occur, starting with fossil fuel carbon being added to the human economic system, then being released into the atmosphere or onto the land surface

7.6 Humans

Finally, we come back to look in the mirror at the last, and perhaps most significantly altered, part of the terrestrial surface, the anthroposphere[19]: the realm of human systems and their effect on climate. As is clear, this goes well beyond just the emissions of CO_2 from burning fossilized carbon fuels. We have already discussed several aspects of human influence: Cropland as a plant functional type and deforestation are examples of human-caused land use change, with subsequent effects on the carbon cycle and regional climate. But human systems also respond to climate, and our economies and underlying social systems are also affected by the climate changes we unwittingly produce. Thus, the anthroposphere is not just a simple flux of "stuff" (whether fossil fuel emissions or acres of cropland), as illustrated by the one-way arrow from the power plant in the carbon cycle schematic of Fig. 7.6. Instead, it is tied to the terrestrial and earth system. Models of how human systems react to climate are now starting to be coupled into terrestrial system components of earth system models. Here, we discuss some of these important feedbacks and how they are simulated.

Figure 7.9 attempts to describe the anthroposphere in the context of terrestrial systems and the climate system. Industrial emissions and emissions from our energy system flow into the atmosphere. But the climate system also forces changes in the energy system: Hotter climates increase demand for energy used in cooling.

---

[19]Also called the anthrosphere. The term also has a companion for a geological epoch, *the Anthrocene*. See Crutzen, P. J. (2002). "Geology of Mankind." *Nature*, *415*(6867): 23.

Changing water availability affects industry and also affects agriculture. Agricultural land (pasture and cropland) has very different surface properties than natural vegetation, which can result in significant differences in evapotranspiration, affecting precipitation, and albedo, affecting surface temperature. Changes in precipitation and temperature in turn feedback on crops: requiring changes to crop types or additional irrigation water if available. All of these feedbacks can be predicted and modeled, with varying degrees of fidelity. Ultimately, human systems like industry, energy, and agriculture respond to price signals from the economic system. The costs of energy and agriculture are affected by the natural environment. This is most obvious with agriculture: Rain and temperatures strongly affect crop yields and the necessity for irrigation. Then crop yields in different regions affect overall prices for crops, the mix of crops, and ultimately the economic system.

Many terrestrial systems models are starting to include complex representations of the physical side of the anthroposphere: agriculture and urban environments. Agriculture is a significant fraction (one-third) of total land area,[20] so it has a large physical effect. The effect of urban areas is significant as well due to significantly different albedo and evaporation characteristics of hard (nonporous) surfaces common in densely built-up areas (roofs and roads). The extent of urban area (less than 0.5 % of the total land area) is small. However, cities have a large impact because of the intensity and magnitude of their emissions. It is also important to simulate urban areas because cities are home to half of the global population. Simulation of urban environments generally starts as a discrete land surface type in a model, but it may evolve to have its own emission characteristics as well.

There are many varieties of economic models. Economic models range from simple supply-and-demand curves (an economic model of a single product in a spreadsheet), to complex models of entire economies or even the global economy. Such macroeconomic or sectoral models typically have similar supply-and-demand curves for different regions of the planet and different economic activities that are interdependent, and can be solved for a solution to all these supply-and-demand equations that end up yielding predictions of economic output and prices. These **economic system models** encompass the Industry, Energy, and Agricultural sectors of the Human Systems box in Fig. 7.9.[21]

Economic models can also be coupled with the climate and environmental factors. Climate factors can alter supply-and-demand curves (crop yields based on weather, for example). The result of these feedbacks is ultimately to alter the trajectory of human systems to react to changes in the environment: If climate changes, crops may not be viable in certain regions, and the economy and society will adapt. Some regions will suffer, but other regions may benefit from increased crop yields in colder climates that have warmed, and this also needs to be factored

---

[20]Data from UN Food and Agriculture Organization (FAO) Statistics Division (FAOSTAT), http:// faostat3.fao.org/.

[21]An economic system model is another name for a macroeconomic model, a model tool designed to simulate a country or region. Many different types, descriptions, and simple models are available on the web.

in. There are basic empirical economic relations that are applied (supply and demand). Then when climate affects human systems, the systems will respond by altering their outputs: including altering their emissions of greenhouse gases. Thus, instead of specifying levels or emissions, such models can attempt to predict emissions by simulating the economic results. They can also attempt to look at the future state of the economy as it co-evolves with climate. When used for assessing future economic states and different economic policy options, these are termed *integrated assessment models.*

7.7 Integrated Assessment Models

Integrated assessment models are generally macroeconomic models that include some linkage to the earth system, and some relationships for the feedbacks in Fig. 7.9 between the physical and human climate systems.[22] Traditionally, such models have a simplified representation of climate (often an energy balance model, or a simple regionally average climate) to represent the physical part of the system. With increases in computing power, however, key parts of these models are increasingly being coupled to full climate system models (see Chap. 8). Essentially, the macroeconomic component is a big series of linked supply-and-demand curves. Supply and demand curves indicate what products are delivered and services produced for demand for different economic sectors. Construction demand is fueled by the need for houses, this causes demand for materials, and the people who receive money for the materials and houses they build then buy cars and houses themselves, etc. These economic equations are coupled to the physical system and is discrete geographically (usually by country, but often with some distribution of effects based on physical distributions of population). There are several challenges to this approach, however.

Humans are unique in that they (sometimes) plan for the future. Coupling of integrated assessment models tries to reflect this. An integrated assessment model would be run every time society wants to adjust for the future. This might be the end of every 10 years of a climate model run. The assessment model might run forward 100 years, and the new trajectory of emissions used for the next 10 years in the climate model. Predicting what will happen in an economy for the next decade includes projecting out the economy for many years and deciding if policies and laws should change, then stepping forward with the altered economy, and then doing it again next decade. This is very different from the physical system, and it makes coupling of economic models and physical models difficult.

Another challenge is inherent to economic simulation itself: The "laws" of economics are merely empirical results. There is no conservation of energy and mass in economies to guide a model. Money and people can be "created" or

---

[22]Parson, E. A., & Fisher-Vanden, A. K. (1997). "Integrated Assessment Models of Global Climate Change." *Annual Review of Energy and the Environment, 22*(1): 589–628.

"destroyed" in the human system at will. The laws of economics work, until they don't, and appropriate economic theory is often contradictory, and melded with politics. A simple example is whether countries with economies in recession (not growing) should borrow money for the government to spend to grow out of the recession, or reduce spending by the government to give people more money. For the latter effect, some economists claim that reducing government revenue (taxes) adds to economic activity, while others claim it does not. The economic debate on this issue is raging in the second decade of the 21st century.

Furthermore, the economic relationships vary over time. The supply-and-demand curve for gasoline will change as technology for both extraction and use changes. Predicting the future for these models is fraught with the problem of being forced to project the past into the future. Usually technological change is treated as incremental improvements/changes to processes or consumption. But technology is not smooth: It is often disruptive. What would an economic simulation of the news and media (or book publishing) business from 1990 look like today, over 20 years later? The 1990 prediction might estimate printing presses would have gotten more efficient, but it would certainly not "predict" the impacts of the Internet or electronic music players, smartphones, and tablets on consumption of books and music. Or consider the example of energy system models, which tend to react from crisis to crisis, and do not see sudden changes well. The rise of hydraulic fracturing technology, which allows fossil fuels to be extracted more economically from different types of geology, has drastically reshaped energy markets and the relative cost of different fuels just in the period 2007–2013. We return to these issues in Section III, on uncertainty.

7.8   Challenges in Terrestrial System Modeling

There are many challenges and complexities in modeling the different parts of the terrestrial system. Some challenges are related to modeling of specific pieces (e.g., land ice), and some are challenges that integrate across the different pieces of the terrestrial system (biogeophysics, hydrology, humans, and nutrient cycles).

7.8.1   Ice Sheet Modeling

Ice sheet models are still developing rapidly. Their development has been spurred by recent observations of significant changes to the ice sheets that raise concerns about changes in global sea level (see Chap. 8). There are several ongoing challenges in developing ice sheet models. First is uncertainty in the complex topography of the base (bottom) of ice sheets. Not surprisingly, often the topography is not fully detailed, as it must be sensed through miles of ice. In addition, there exist uncertainties in some of the dynamic processes that occur, such as the water that lubricates glacial flow. So the problem becomes similar to the complexity and heterogeneity of the land surface, but now the whole surface model is in slow motion. Another issue is subglacial melting from ice shelves. Melting from contact with ocean water underneath floating ice sheets or shelves is strongly dependent on ocean circulation. This problem is critical for Antarctica.

Land ice models must also be run with relatively long time steps for thousands (or hundreds of thousands) of years, making the simulation of ice sheets in the coupled climate system very challenging. Because of the complex topography under the ice sheets (and limited area), they are often run with small grid sizes (high horizontal resolution). These models are challenged also by the lack of key data, such as the detailed topography under the ice sheet, and especially limited observations of surrounding oceans underneath thick floating ice shelves. Thus dynamic simulation of ice sheet processes is a challenging task.

7.8.2 Surface Albedo Feedback

The albedo feedback hinges on sudden changes at the terrestrial surface. Albedo is the "absorption fraction" of the surface and depends on the color: Dark surfaces such as the oceans, or a dark green forest, absorb more light (and have a high albedo, close to 1). Light surfaces, such as snow, ice, and light, bare soil as found in deserts, have a low albedo (close to zero). Albedo can refer to any wavelength, but here we refer to solar wavelengths (visible light from the sun). Reflecting solar energy from light surfaces tends to cool; absorption by dark surfaces warms.

This make a classic positive feedback: If sea ice over the dark ocean or snow over darker vegetation melts, then the albedo goes up, the absorption of energy goes up, and the temperature goes up, melting more snow and ice and exposing more dark ocean. Conversely, if the temperature drops, ice and snow expand, reflecting more light and cooling the surface, resulting in more snow and ice. Note that a positive feedback amplifies both ways; it amplifies cooling and warming. A negative feedback damps both ways, causing changes to be minimized.

The connotation of feedbacks in terms of climate change is actually the *opposite* of common usage: Negative feedbacks are usually "good" (they damp changes), whereas positive feedbacks cause larger changes (bad, especially if you are a polar bear). When you receive negative feedback from your boss, however, it is usually *not* good. The snow-ice albedo feedback is a big amplifier of climate changes in snow- and ice-covered regions, and it is a reason why the Arctic has warmed more than other regions recently. It is also a mechanism that naturally comes into play during ice age cycles, as advancing glaciers and sea ice cool the planet. The feedback depends on exactly how much snow and ice there is: If the snow is too thick to melt at some point in the annual cycle, then the albedo doesn't change much for a given heat input. Also, if there is little snow left to melt, there is not much temperature change with more heat. This means that the ice-albedo feedback contribution to the climate sensitivity is variable with the current climate state. As a practical matter, this makes looking into the past for paleoclimate records of previous ice ages not that useful (and potentially misleading) for understanding the present sensitivity of the system.

7.8.3 Carbon Feedback

In Chap. 5, we discussed cloud feedbacks, and in this chapter we discussed surface feedbacks with ice and snow. Cloud feedbacks are "fast" (minutes to hours) and ice and snow feedbacks are "slow" (decades to centuries for ice sheets). There are also a spectrum of slow feedbacks related to the cycling of carbon in the climate system. The simple example is the land carbon in soils and plants. Changing the level of CO_2 makes plants grow more efficiently. With more CO_2 in the atmosphere, plants open their pores less to let in CO_2, which reduces water loss and makes them more efficient. An analogy would be what happens to humans going from higher to lower altitudes: As oxygen increases, breathing is more efficient (though we usually experience this in reverse when we go to higher altitudes).

So what happens if CO_2 increases (if all else is equal, which is a big "if")? Plants would tend to grow more, and this would increase their CO_2 uptake, reducing atmospheric CO_2 (a negative feedback). This assumes that plants are "limited" by CO_2 and not by water and nutrients. It may not work if nutrients or water are limited. For a human, more oxygen will make you more efficient at breathing, but you still need enough food and water. In addition, warmer temperatures may increase the decay of plant material (e.g., dead leaves) that returns CO_2 to the atmosphere and leaves less in soils. There are many feedbacks with the land surface that, rather than changing the energy budget, directly change the partitioning of carbon between land (or even the ocean) and the atmosphere. The change in CO_2 in the atmosphere changes the energy budget.

These carbon feedbacks may be important on long timescales and would modulate the fast feedbacks in the atmosphere, making them critical to understand for long-term climate change. In addition, storage of carbon in the ocean and ocean ecosystems can also affect atmospheric CO_2, and the global carbon cycle.

7.9 Applications: Wolf and Moose Ecosystem, Isle Royale National Park

This case study demonstrates the methodology of participatory scenario planning for a terrestrial system and how consideration of a specific application defines the role of uncertainty (a point made again in Chaps. 11 and 12). Isle Royale is a small national park in Lake Superior, the largest of the U.S.-Canadian Great Lakes. A unique and valued attribute of the park is a precarious balance between the wolf and moose in the park: a predator-prey ecosystem. The existence of the ecosystem with both animals is generally attributed to sporadic formation of an ice bridge between the island and the Canadian mainland. With the dependence on an ice bridge, there is an explicit relationship between sustaining the populations of these species and maintaining a diverse gene pool by communication with the mainland. This communication is dependent on the climate that creates this ice bridge.

This case study involves a scenario-planning process where plausible, not probable, futures are developed to facilitate investigation of management decisions and develop preparedness.[23] Physical climate projections of temperature and precipitation from climate models have been used to describe projected changes in the environment for a time range in which management decisions are consequential (decades). The focus is on the projection with the "least change" from the present. Though more extreme projections are considered, the least change projection lies at the foundation of the formation of scenarios. The scenarios are formed through a participatory process where, for example, an extreme event (e.g., a wind storm) is conjectured with ecological consequences (e.g., trees blow down). Multiple scenarios are considered, with the responses framed by management priorities, which might include conservation requirements, wilderness management, infrastructure, or visitor experience.

Isle Royale is a small park (about 250 square miles, or 650 km^2), smaller than the resolution of global land models used in climate models. Though large, Lake Superior is not represented with fidelity in global climate models, often treated as a land-surface type. The local, lake-influenced weather processes, responsible for the park's climate, are not represented well in current climate models. Moreover, physical processes such as summertime convective precipitation have large regional biases in models. Therefore, there are substantial barriers to direct, credible application of climate model projections.

In addition to the structural shortcomings of climate models, parameters important to the park, and in particular to the wolf and moose ecosystem, are not directly simulated. An overt example is lake ice, fundamental to the existence of the wolf and moose populations by connecting the island to the mainland. Other parameters include snow cover and winter melt, which directly influence access to moose browsing habitat (food). Decadal trends in observed lake ice cover indicate up to 70 % reductions, with this trend interrupted by extremely cold winters (e.g., 2013–2014) with high amounts of lake ice. This brings attention to variability and in this instance focuses the discussion on questions of changes in variability hypothesized as a possible response to long term changes in the high Arctic.[24]

---

[23]Details and results of this case study can be found in "Using Climate Change Scenarios to Explore Management at Isle Royale National Park," http://www.nps.gov/isro/learn/nature/using-climate-change-scenarios-to-explore-mangement-at-isle-royale-national-park.htm.

[24]The possibility that changes in the Arctic might have strong influence on mid-latitudes was proposed by Francis, J. A., & Vavrus, S. J. (2012). "Evidence Linking Arctic Amplification to Extreme Weather in Mid-Latitudes." *Geophysical Research Letters, 39*: L06801. doi:10.1029/2012GL051000. This is an active research area.

Climate observations and climate projections provide a background for discussion. Essential in the application was evaluation of local weather processes and whether or not models represented these processes. This helped to develop trust of the expert guidance to interpret model information. Attention is naturally drawn to recent extreme events and whether or not these extremes are consistent with projections, for example, more precipitation occurring in extreme events. Warm and dry spells in the winter and spring that alter greening of forests, followed by damaging cold, is another example. Convolution of climate, extreme events, and ecological responses sit at the foundation of plausible futures. For example, if there is large-scale disruption of forests by drought, fire, or wind that leads to the death of many trees, then the future forest will be recovering in a much different mean climate than in which it originally evolved.

Since Isle Royale's forests are at the southern extent of the subarctic (or boreal) forest, and that extent may well move northward in a warmer climate, it is unlikely that a boreal forest disrupted by drought, fire, or wind will be regenerated. Given the importance of specific tree species to moose food supply, this would be a negative indicator for moose populations. Evaluation of the combined influence of climate drivers was largely negative for maintaining the wolf-moose ecosystem. Though climate change is only part of the portfolio of factors in the decision-making package, it demonstrates that in the future it will be even more difficult to sustain this precarious ecosystem. A driving conclusion from this exercise is the need to plan for best possible futures rather than manage toward preservation or conservation of the past.

7.10   Summary

Modeling the earth's surface means modeling a complex set of coupled terrestrial systems. The surface fluxes that occur in the climate system are strongly affected by key properties of the surface: Water fluxes are affected by transpiration from plants. The presence of water is also important for moving energy around and releasing it as latent heat (analogous to the role of water in the atmosphere). This is very important in semiarid regions with dry soils. Transpiration of water from plants is an important part of surface processes. And since plants and ecosystems are dynamic and respond to climate, representing different plant types and the ecosystems that support them is critical. Furthermore, the growth and decay of plants in ecosystems depends on critical nutrients, such as carbon and nitrogen. Carbon is the common lifeblood of the earth system, changing forms from the solid earth and sediments, to biological tissue on land, in soils and in the ocean, to a greenhouse gas in the atmosphere. Understanding carbon couples terrestrial systems to climate as well. These systems include the cryosphere (snow and ice) and the anthroposphere (agricultural land, urban areas).

Modeling terrestrial systems involves several components. The biogeophysics of the system is described by a model of energy and water flows, including the absorption and emission of radiation. The hydrology of the land and the terrestrial water cycle is also simulated: Precipitation is input; evaporation, transpiration, and storage in soil moisture occur; and the remainder becomes runoff.

   Terrestrial systems generally include a description of the type of ecosystems (plant types) on the surface and soil properties in the subsurface, often in detailed small-scale tiles. The descriptions of the plant types are typically based on climate effects and are not necessarily considered a "detailed" description by an ecologist. Descriptions of plant types represent the effects of ecosystems, or a population of plants, not individual plants. This is similar to parameterizations of clouds in the atmosphere designed to represent a distribution of clouds and their effects, not a single cloud. Different ecosystems have very different properties (height, leaf area, root depth, and transpiration) that affect surface fluxes. The ecosystem descriptions can be dynamic and evolve over time.

   Key nutrient cycles, usually carbon and then nitrogen, are often added to land surface models to improve the ability to simulate changes in terrestrial carbon budgets. Changes to the land surface cycling of carbon can alter CO_2 storage and emission.

   Ice sheets and snow are an important land cover type for altering albedo and solar energy absorption at the surface. And ice sheets are important for storing water that affects global sea level. There are many complex and incompletely observed processes that determine the balance of ice sheets between accumulation, melting and flow. The Antarctic and Greenland ice sheets have different characteristics and critical uncertainties. The Antarctic ice sheet is sensitive to ocean processes beneath ice shelves and at the edges. The Greenland ice sheet is sensitive to melting on the surface and lubrication at the base.

   Finally, many of these land properties are affected by human systems, and these human systems are tightly coupled to the climate system in two-way interactions between the climate system and human industrial, energy, and agricultural uses. Macroeconomic models can simulate these human systems and can be coupled to physical climate models to try to provide possible future "predictions" or scenarios of the co-evolution of natural and human terrestrial systems.

Key Points

- Plants play a large role in climate by moving moisture through transpiration.
- Land surface models represent soil and soil water, and many plant types.
- Nutrient cycles, like carbon and nitrogen, are important at longer timescales in the climate system.
- There are many challenges in modeling ice sheets in climate models.
- Human system models (economic models) can also be coupled to the climate system.

Chapter 8
Bringing the System Together: Coupling and Complexity

Chapters 5–7 focused extensively on what parts of the earth system are being modeled. This chapter brings together the different components of the climate system and discusses how the component models we have described are coupled together to represent the earth system. This chapter is more about the mechanics of models, and how we couple the pieces together. Different components (atmosphere, ocean, ice, terrestrial systems) each have their own complexities, and their own challenges for modeling. Different types of coupled models can be constructed, representing regions, the whole planet, or even focusing on human systems.

Some specific features in the climate system are really the product not just of one component, but of the interaction of different components. One example might be the variability in tropical ocean surface temperatures in the Pacific, called El Niño. Every few years the eastern Pacific water warms up, in an oscillation that has large impacts on global weather patterns. Treating these patterns or modes of variability properly is necessary to properly represent climate.

There are different ways of running different models (e.g., regional or global, full ocean or mixed-layer ocean only). Some important aspects of the coupled system are obvious. For example, the amount and distribution of precipitation from the atmosphere strongly impacts the terrestrial surface. Other aspects of coupling are more complex. A number of them have been described, and we focus here on a few more interactions and challenges, especially those interactions that result in strong feedbacks to the climate system by ultimately affecting the energy in the climate system. We also examine some features of the system, such as the global sea level, that are set by interactions of the ocean, ice, and atmosphere.

8.1 Types of Coupled Models

The traditional models we have described are general circulation models (GCMs), which are dynamical system models of the atmosphere, ocean, and other pieces of the system (e.g., sea ice, terrestrial systems). These are global and three-dimensional models that are designed to represent key earth system processes. But these are not the only kind of climate model.

© The Author(s) 2016   139
A. Gettelman and R.B. Rood, *Demystifying Climate Models*,
Earth Systems Data and Models 2, DOI 10.1007/978-3-662-48959-8_8

Increasingly, climate models are being extended to incorporate other types of models that are of direct relevance to society, such as ocean-wave models, hydrologic models, ecosystem models, and policy-relevant models. The complexity represented by these models is daunting. Some researchers develop local models of high complexity and limited temporal and spatial application. Coupling with these models is sometimes "one-way" in the sense that the global climate model provides information to, but does not receive information from, the detailed process model. Sometimes the coupling is "two-way" so the effect of the detailed model changes (or feeds back on) the global model. These detailed models can be classified as *impacts models.*[1]

An example of an impact model related to a physical process is a global ocean-wave model, which would be expected to have a "two-way" coupling. That is, the climate model surface winds will determine the wave field and the wave field will affect surface characteristics, for example, water and salt exchanges, as well as surface drag. The output of the wave model will influence the climate model. There are also models of economic and technological relationships and responses to climate and climate change, which is a type of integrated assessment model. Below, we focus on a small number of coupled models that have sufficient maturity that they have broad exposure to practitioners interested in applying model projections.

8.1.1 Regional Models

Regional climate models[2] are also three-dimensional dynamical systems models, but with smaller domains. As a result. they often contain often more detailed processes. They are run with boundary conditions from observations or from GCMs (other models) when no observations are available (i.e., in the future). Regional climate models are often used to generate high-resolution and high-frequency weather statistics to understand how broad-scale climate change alters weather patterns. Examples of regions might be Western Europe, the Arctic, or the continental United States. The atmosphere models used are often models that are used for weather prediction, that resolve the scales of weather systems from 3 to 125 miles (5–200 km). These are often called *mesoscale* models (*meso* = medium). These are an effective way to "downscale" the large-scale general circulation (and potential changes) to generate better weather statistics by driving dynamical models at high resolution with the output from a global model (see box in next section).

Why is higher resolution (smaller grid boxes) better? One strong driver for local weather and climate is the physical environment, especially topography (landforms and changes in elevation) that affects both the atmosphere and the ocean. At lower

---

[1]For more information on such models, see the International Environmental Modeling and Software Society, http://www.iemss.org/society/index.php/scope.

[2]Rummukainen, M. (2010). "State-of-the-Art With Regional Climate Models." *Wiley Interdisciplinary Reviews: Climate Change, 1*(1): 82–96.

resolution, the topography is just not represented correctly because a large grid box at low resolution can have only one elevation: the average of perhaps many mountains or valleys. It is thus hard to represent the climate of California, Chile, Colorado, or Switzerland, without resolving mountains. One way to think about resolution generally is that in regions where there is a lot of variability in any quantity, like elevation (though it could be surface albedo or even atmospheric water vapor), increasing the resolution allows more of the variability to be represented explicitly, and to force the climate directly. From the practitioner's perspective, higher resolution represents features at a geographic scale that is more intuitively relevant, such as more realistic and detailed coastlines and the built environment of cities.

8.1.2 Statistical Models and Downscaling

GCMs and regional climate models are dynamical models. They use the equations of motion and thermodynamics to determine the rates of change of physical quantities (e.g., water vapor, temperature or heat, cloud water, carbon). There are also statistical models of climate. **Statistical models of climate** take observations from the past and try to predict the future with various forms of regression or correlation analysis: fitting past data on temperature and precipitation for example, to a function that is used as a predictor of the future. Usually this is done for **downscaling**, fitting temperature at one point to a larger scale temperature or flow pattern that can be predicted by a dynamical model.

> **Regional Climate Modeling: Downscaling**
> Use of a regional model at high resolution is an example of at type of analysis called **downscaling**.[3] Downscaling uses finer-resolution information to improve the results of a coarse-scale model. It is effective especially when the improvement in resolution affects the simulation because of small-scale features at the surface: as is the case in regions of varied topography (mountains or coastlines). Use of a regional or local area model is known as **dynamical downscaling**. Downscaling can also be **statistical downscaling**. For example, if you want to know the temperature high on a mountain, you could develop a statistical relationship between the temperature for the whole region and this particular point based on observations over the past 50 years, and then adjust climate model output for the region in the same way to get a simulated record at that station that takes into account the unique local features (high altitude). Both methods of downscaling would be particularly useful for representing climate in regions of variable topography: either at high altitude,

---

[3]Wilby, R. L., Wigley, T. M. L., Conway, D., Jones, P. D., Hewitson, B. C., Main, J., & Wilks, D. S. (1998). "Statistical Downscaling of General Circulation Model Output: A Comparison of Methods." *Water Resources Research, 34*(11): 2995–3008.

or in mountain valleys and around mountain ranges. Downscaling works best when we have a good physical understanding of the variations, like temperature, than for things like precipitation that are dependent on small-scale atmospheric processes. In mountainous regions, precipitation is also strongly dependent on the flow of air over mountains and its direction.

Downscaling with statistics works only if (a) the statistics are well described, (b) the relationships do not change, and (c) the data do not need to be extrapolated. Statistical models can work for weather events, largely because we have a lot of statistics on daily weather at a particular place, and small-scale details are dependent on large-scale patterns. Note that weather forecasts are *not* done with statistical models: Weather prediction models have more skill than statistical models. But if you want to understand what the climate of a particular place is when all you have is the large-scale flow (from a dynamical GCM), then basing the results on the past may be accurate.

Let's say we have 60 years of daily weather data. Then for any day in a particular month, there are 1800 samples of days (30 days × 60 years) from which to build a statistical model at one place for temperature, or precipitation, or stream flow. The assumption of statistical downscaling is that these samples represent all possible states. The problem in doing this for the future is that if the climate is changing, then by definition the probabilities and statistics are changing. Recall Fig. 1.1, showing the shifting probability distribution function (PDF) of climate. The distribution that defines climate is changing, so the statistics change, and there are extremes that are not in the past statistics (so extrapolation would be necessary).

There exist inherent dangers in statistical downscaling, but the results may be better than just using large-scale dynamical models that miss important local effects. This is true, for example, for places where the climate is affected by topography that is not well resolved by global models with a resolution of 16–65 miles (25–100 km). Practically, the resolved resolution is at least four times the grid resolution, so this means features smaller than about 65–200 miles (100–400 km).[4] Examples would include islands that might not be resolved by large-scale models, and peaks and valleys in mountain ranges that have steep gradients. A very good example of the combination of both would be the local climate on the island of Hawaii: It is dependent strongly on your location on the island, the elevation, and the direction of the prevailing large-scale wind. Here a statistical model of climate may be helpful: If you know the large-scale wind at a given location, you can probably represent the temperature halfway up a mountain pretty well. But in general, even here, dynamical downscaling using regional climate models may be preferred.

---

[4]Kent, J., Jablonowski, C., Whitehead, J. P., & Rood, R. B. (2014). "Determining the Effective Resolution of Advection Schemes. Part II: Numerical Testing." *Journal of Computational Physics,* *278*: 497–508.

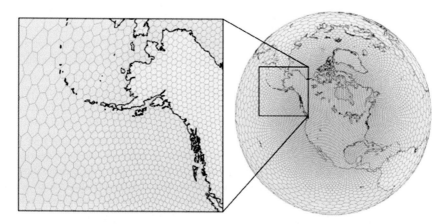

Fig. 8.1 An example of a variable resolution grid from the model for prediction across scales (*MPAS*). The grid gets finer over the continental United States using a grid made up of hexagons. *Source* http://earthsystemcog.org/projects/dcmip-2012/mpas

An active area of work right now is using increased computer power to run GCMs at the resolution of regional models, to have global models capable of producing regional climate statistics. The goal is to focus on representing extremes of weather events (the tails of the PDF). Regional models have typically been run with grid sizes of 3–17 miles (5–25 km). Global model grids have been 30–140 miles (50–200 km). Global models in the range of 5–25 km are now running experimentally, and some are even capable of running with variable resolution grids, where the resolution is finer in one region of the planet than in other regions. An example is shown in Fig. 8.1 (also shown in Chap. 4, Fig. 4.2), where a complex hexagonal grid has fine resolution over North America (10–20 km) and coarse resolution (100–200 km) elsewhere. The resolution inside of the fine mesh is comparable to a regional model. The total number of grid points for which the equations of state must be solved is not much more than a regional model, because the outer grid is sparse. There are difficulties in dealing with the different resolutions for representing both the motion of air and the representation of clouds at different resolutions.

8.1.3 Integrated Assessment Models

Models relevant to society and decision making represent another class of models that might be coupled with climate models. This class of models might be focused on local adaptation or global mitigation. Of specific note, here, are integrated assessment models (IAMs).[5] As noted in Chap. 7, these models are focused on the anthropo-sphere, the human part of the terrestrial and climate systems. These models have a societal component that resembles a macroeconomic model (see Sect. 8.6 for an application of such models). Macroeconomic models are dynamical systems models in which the grid points are locations like countries, or groups of countries, with a single economic system. There are still linkages between the grid points (trade flows). Then this economic model is coupled in some way to a simplified representation of the physical system. The land surface component (natural ecosystems) could have a grid corresponding to the economic model, and the ocean model might simply be a basin model. Sometimes the atmosphere is represented as a one-dimensional energy balance model at each grid point, sometimes the world is a simplified circulation model with low resolution, sometimes a series of statistical relationships.

Integrated assessment models can have an economic component coupled to a full climate model, but generally the climate system is substantially simplified. The goal of an IAM is to determine the broad-scale climate impact of the economic system, project that forward, and then allow the social system to adjust, perhaps changing itself in the process (e.g., new policies), then projecting the climate again. The result is a co-evolution of climate and society.

A number of efforts are being conducted to couple economic models to full climate models. One hybrid would be a climate model with an anthroposphere in the terrestrial system. These models are designed to generate self-fulfilling prophecies: Rules are set for policies that react and change as the climate chan-ges. The policies adjust and then adjust the climate, and so on. The goal is to estimate economic policies, and the costs of the outcomes. The complexity of these models is that IAMs have very different time horizons than dynamical models. Typically, an economic model would be run for 100 years forward with a simple "projection." That projection would affect policies in the model, and then the physical and economic system would run 1 year forward. Then the IAM would run an adjusted scenario again for 100 years.

8.2 Coupling Models Together: Common Threads

For GCMs, there is quite a bit of complexity in coupling the different components we detailed in Chaps. 5–7. The creation of a coupled system model proceeds by improving each component (e.g., atmosphere, terrestrial surface, ocean), as well as trying to put the components together and make them work together. Improving each component means improving the representation (parameterization) of each process, as well as improving the coupled system as a whole. Ultimately, the processes that provide the rates of change for the state of the system then affect the

---

[5]"Integrated Assessment Modeling: 10 Things to Know," http://sedac.ciesin.columbia.edu/mva/iamcc.tg/mva-questions.html.

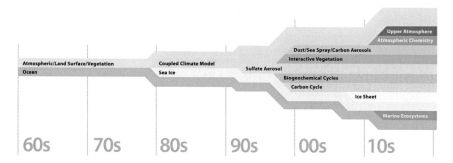

Fig. 8.2 Schematic of components. Evolution of the parts of the earth system treated in climate models over time. Figure courtesy of UCAR (same as Fig. 4.8)

whole system. For example, clouds affect the radiation absorbed and emitted as well as precipitation, and this then strongly affects the underlying land or ocean surface.

The coupling of different components is achieved in a coupling layer that is a bit like a clearing-house for financial trades. Changes to financial positions come in (e.g., stock trades). The changes are settled between the accounts of the buyer and the seller, their accounts updated, and the stock changes hands. In a similar way, each component model has changes in their state (e.g., precipitation from the atmosphere hitting the surface). Mass and energy budgets need to be conserved in the same way that stock certificates and money have to be accounted for. And, like a modern financial market, there are lots of traders (one for every grid box), trading different stocks (the different states of the system, like mass of water, or carbon or the energy in a region). And all this happens over and over again. There is a regulator that ensures all this runs smoothly. In a model, the regulator checks the mass and energy to ensure conservation. Fortunately, the physical equations in a climate model are usually not prone to the self-criticality ("crashes") of financial markets. That is because there are lots of negative feedbacks in the system that damp out the changes. For example, if the temperature rises because there is a lot of sunlight, then more energy is radiated to space. Or warmer temperatures might cause upward air motion, condensation, and the formation of clouds, shading the surface.

Coupling the components we have described began slowly, as illustrated in Fig. 8.2. Coupling, of course, occurs at the interfaces of the components, and this mostly means coupling at the surface of the earth. As circulation models of the atmosphere and ocean were developed, the atmosphere generally had a specified land surface. Initially, each model received specified surface exchanges of energy and heat based on observations. In the late 1960s,[6] about 10 years after the first models were developed, the atmosphere and ocean were coupled together, with the

---

[6]Manabe, S., & Bryan, K. (1969). "Climate Calculations With a Combined Ocean-Atmosphere Model." *Journal of the Atmospheric Sciences, 26*(4): 786–789.

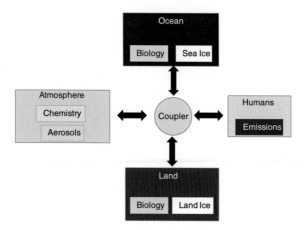

Fig. 8.3 Coupled climate model. Schematic of the component models and subcomponents of a climate model program. The coupler code ties together different spheres (ocean, atmosphere, land, biosphere, and anthroposphere) that then contain smaller parts (like aerosols, chemistry, or sea ice)

first real geographically resolved simulations in the mid-1970s.[7] The coupling was fairly crude, with the models not really run together. The ocean model was run to provide forcing for the atmosphere, and then the atmosphere model was run to provide forcing for the ocean, and so on. The ocean models were usually not models with a deep ocean circulation. Figure 8.2 illustrates the timing of this evolution. Sea-ice models were added to ocean components in the 1980s. In the 1990s, detail began to be added to the land surface, and the first simulations of some of the biogeochemical cycles were attempted.

Now, the coupling of the model components is more synchronous, with atmosphere, land, and ocean models running with the same time resolution: several steps forward at a time, with coupling occurring each day, or several times a day. The coupling of the components illustrated in Chap. 4 (Fig. 4.9) is illustrated again here in Fig. 8.3. The different model components (many with different submodels for processes) all are coupled together, and are integrated forward (run) at the same time.

One of the major problems with this approach has been that errors in the different processes and different components resulted in consistent and significant errors in the surface fluxes or exchanges passed to other models. Think of having the wrong cloud cover in a region: Significant errors would mean too much or not enough solar energy reaching the surface and going into the land or atmosphere. An analogy would be an imbalance in financial flow. If a stock trader consistently took

[7]Manabe, S., Bryan, K., & Spelman, M. J. (1975). "A Global Ocean-Atmosphere Climate Model. Part I. The Atmospheric Circulation." *Journal of Physical Oceanography, 5*(1): 3–29. See also Bryan, K., Manabe, S., & Pacanowski, R. C. (1975). "A Global Ocean-Atmosphere Climate Model. Part II. The Oceanic Circulation." *Journal of Physical Oceanography, 5*(1): 30–46.

part of the trade proceeds, and did not return it to the client, the money flows returned would be too small to keep accounts in balance. Up until about the year 2000, many climate models adjusted their coupling to correct for systematic errors. A model was run uncoupled, and the biases compared to observations were adjusted away by a fixed "flux adjustment" of heat and/or moisture. Of course, flux adjustment assumes that the biases are systematic, and that these are constant. Like statistical models, this adjustment works if the climate changes are assumed to be small perturbations of a basic state.

Fortunately, the use of flux adjustments has largely been eliminated in most coupled models. The removal of flux adjustment has occurred because better representations of the components of the system have smaller systematic errors. Flux adjustment is like correcting the steering of a vehicle for a lack of alignment in the wheels. With better alignment, it is no longer necessary. Another advance that has helped is to integrate climate models for longer periods of time with steady forcing applied, usually representing a pre-industrial state of the 19th century, before carbon dioxide (CO_2) levels began to rise. This allows all the components to come into balance (equilibrium) with each other.

8.3 Key Interactions in Climate Models

Once components are coupled, there are a number of key interactions that are critical for the earth's climate. We have talked about critical processes in each component: water or latent heat in the atmosphere and land as a mechanism of heat transport, salinity in the oceans, transpiration on land. Let's look at some different couplings of processes in the system and how they are simulated. Many of these are really feedbacks that cause one component to affect another.

8.3.1 Intermixing of the Feedback Loops

All of these feedbacks are playing out in the climate system at the same time. The intermixing of the feedback loops makes simulations incredibly complex and is one of the reasons we turn to finite element models constrained by energy and mass balance for simulation. For example, it is the change in forcing from the sun as the earth's orbit changed that likely ended the last ice age: warming the Northern Hemisphere. This would be amplified by the ice albedo feedback from melting land glaciers, but it also may have changed the ocean carbon cycle, resulting in more CO_2 getting into the atmosphere, and accelerating the warming trend, which melted more snow and ice, and so on. But these feedbacks would work very differently now, because of a different ocean circulation and a different distribution of ice sheets on land. There are also feedbacks associated with biological cycles of methane (natural gas, CH_4), another greenhouse gas.

So positive and negative feedbacks act on the earth system, regulating and changing the response of climate to forcing. On the scale of human systems (a century or so), cloud feedbacks are critical. Clouds respond quickly to changes in their local environment, such as altered surface temperature or altered temperature profiles, and this environment is affected by climate change. Cloud feedbacks are thought to be positive in the current climate (warming makes clouds get thinner and/or less extensive, and this reduces the cooling effect of clouds). Ice- and snow-related albedo feedbacks are important at high latitudes and are probably large now, as ice and snow disappear rapidly.

Uncertainties in climate feedbacks drive our uncertainty about the sensitivity of the current climate. The larger the sum of the feedbacks, the greater the climate response to a given forcing. A larger positive albedo feedback means a higher surface temperature for a given forcing. Since we can estimate the radiative forcing for increasing amounts of CO_2, the climate response is due to the combination of feedbacks. Longer-term feedbacks with the carbon cycle make that trajectory even more uncertain as we extend to centuries, because now the change in climate will affect how much CO_2 is in the atmosphere, which alters the radiative forcing. Cloud feedbacks are confined to the atmosphere. But there are other feedbacks like carbon cycle feedbacks that cross boundaries of components of the climate system, and can be understood only with coupled climate models.

8.3.2 Water Feedbacks

Some of the most important couplings have to do with water. The most obvious is the flow of water through the atmosphere (as clouds and precipitation) to the land surface (soil moisture, runoff, lakes and rivers), to the oceans (inflow from rivers and glaciers), and then back into the atmosphere through evaporation. This is the hydrologic cycle, and it pervades every part of the climate system. Because of the complexity of this system, significant processes are still either missing or uncertain. Recently, terrestrial systems have sought to include detailed models of runoff and transport. Atmosphere models still struggle with representations of clouds and humidity. And observations of the hydrologic cycle are still uncertain, so that there is limited quantitative information to constrain the hydrologic cycle in some regions.

Water is fundamental, not just for moisture as precipitation, but for changing the energy and physical state of the system, often through changes of phase. Water is one of the few substances that can be found naturally on the earth as a gas, liquid, or solid. The changes of phase can have interesting consequences: Changes from vapor to liquid or solid in the atmosphere (the formation of a cloud) suddenly have large impacts on the solar radiation at the surface (one feels immediately cooler when the sun goes behind a cloud). Formation of sea ice expels salt, making the remaining ocean water denser. Water is also an important contributor to the energy budget through latent energy: the energy needed to evaporate water back to vapor, or the energy released during condensation. Latent heat energy in water is an efficient mechanism to move heat around as water vapor, and cloud systems move with the general circulation. Plants as well as animals use evaporative cooling (sweat in animals, transpiration in plants) for cooling. Water vapor is also released by plants as a side effect of photosynthesis.

Also important at the surface, and related to water, is the coupling of soil moisture and evapotranspiration (evaporation including evaporation from plants) with precipitation.[8] Increases in soil moisture increase evapotranspiration, at least in semiarid environments: Wet environments are energy limited (there is plenty of water to evaporate, but not enough energy for it), and arid environments are too dry (there is not enough water). But in semiarid regions, increases in soil moisture increase evapotranspiration, which reduces soil moisture, thus shutting off the coupling (a negative feedback). But the increased moisture in the atmosphere may lead to more precipitation, starting the cycle again.

8.3.3 Albedo Feedbacks

Surface albedo feedbacks allow more heat to be absorbed at the surface, but there are special feedbacks with sea ice.[9] Melting of sea ice changes the albedo of the surface to a dark ocean, but it also changes the surface exchanges of heat and moisture, as it is easier to evaporate water from the ocean than to sublimate (transfer from the solid to vapor phase without liquid) from ice. The result also impacts the atmosphere, and the extra fluxes or exchanges of heat and moisture can either reduce clouds if heating dominates or increase them if moisture dominates. Much depends on how the surface and the atmospheric boundary layer above it are coupled. The results are critical for assessing the feedbacks and the energy budget at high latitudes. These couplings are inherently present in models, but only in models with active and detailed ocean and ice components, combined with a good representation of Arctic clouds (sometimes ice, sometimes both ice and liquid). Needing so many processes, along with sparse observations, makes this coupling hard to reproduce. Complicating the modeling of the Arctic is limited observations. There are now only 5–10 years of high-frequency and high-quality satellite cloud observations in the Arctic after limited long-term records from individual sites.[10]

---

[8]Seneviratne, S. I., Corti, T., Davin, E. L., Hirschi, M., Jaeger, E. B., Lehner, I., et al. (2010). "Investigating Soil Moisture–Climate Interactions in a Changing Climate: A Review." *Earth-Science Reviews, 99*(3–4): 125–161.

[9]Hall, A. (2004). "The Role of Surface Albedo Feedback in Climate." *Journal of Climate, 17*: 1550–1568. doi:10.1175/1520-0442.

[10]For more on the Arctic and climate, see Arctic Climate Impact Assessment, http://www.acia.uaf.edu.

8.3.4   Ocean Feedbacks

There is also coupling between the atmospheric winds and the ocean circulation. Particularly around Antarctica, the wind flowing off the continent pushes ice off shore, forming leads (open water). The open water cools due to the exposure to the cold atmosphere and forms more ice. The salinity increases from salt expelled from ice that is forming. This makes the remaining water denser. The cold and salty (dense) water sinks. Antarctic bottom water is some of the coldest and densest water in the oceans, and it flows into the Atlantic and Indian Oceans (the formation regions are not on the Pacific side of the continent). Wind stress also affects other areas of the ocean, such as along ocean boundaries and along the equator. These are just two examples of the coupling of the atmosphere and ocean. If the atmospheric circulation changes, it alters the ocean circulation. The same may also happen in reverse, as the ocean circulations move heat around. Changes in the North Atlantic western boundary current (the Gulf Stream) will change the heat transport into the North Atlantic.[11]

8.3.5   Sea-Level Change

The change in sea level experienced in any location results from several factors, many of which must be simulated to properly understand how sea level may vary. First, and rather obviously, the sea level depends on the amount of water in the ocean. As ice sheets melt (or grow), they change the total water in the ocean. This changes the sea level. During the last ice age 20,000 years ago, the sea level was 400 ft (120 m) lower.[12] But there are other factors that change sea level, and many of them are local. Since the density of water can change, its volume changes, too. If the ocean warms up and warmer water is less dense, it expands and raises the sea level as well.

These processes affect global sea level. But the sea level does not change the same everywhere on the earth. Why not? Small, localized changes in the height of the land surface occur over time due to the shifting of the land surface itself. The most important of these changes is the result of large ice sheets having been on many regions of the Northern Hemisphere until 10,000 years ago. The weight of these ice sheets pushed down the land surface. When the ice sheets receded, the land surface began to recover and rise. This rebound (technically, **isostatic rebound**) reduces any impact of sea-level rise from increasing volume of the ocean (due to more water or warmer and less dense water) because the land is rising locally.

---

[11]For a review of the ocean circulation and the North Atlantic in Climate, see Vallis, G. K. (2011). *Climate and the Oceans*. Princeton, NJ: Princeton University Press.

[12]Fairbanks, R. G. (1989). "A 17,000-Year Glacio-Eustatic Sea Level Record: Influence of Glacial Melting Rates on the Younger Dryas Event and Deep-Ocean Circulation." *Nature, 342*(6250): 637–642.

In addition, there is an important "tilt" to ocean heights caused by the motion of the ocean. When water moves (due to wind stress), the forces on it cause variations in pressure across the flow. This pressure difference allows water to stack up where there is lower pressure, tilting the ocean surface slightly. When the circulation changes, the tilt of the ocean changes. This is important locally. For instance, the Gulf Stream current off the east coast of the United States causes a tilt of the ocean surface to lower the ocean along the coast. If the circulation weakens, the ocean tilt will relax, causing the coastal sea level to rise even more in regions near the circulation.

8.4   Coupled Modes of Climate Variability

The couplings described in the preceding sections are critical emergent properties of the climate system that depend on other components. Their coupling can change the mean state of the system. Some couplings, however, give rise to different patterns and timescales of variability. These couplings have significant impacts on weather and climate scales. Here, we discuss small and fast timescale processes of tropical cyclones (hurricanes), seasonal monsoon circulations, tropical oscillations of El Niño, drought and precipitation linkages, and the enhanced efficiency of CO_2 uptake at higher CO_2 concentrations.

8.4.1   Tropical Cyclones

Tropical cyclones, known as hurricanes in the Atlantic and as typhoons in the western Pacific, are a critical and extreme weather event with climate implications. Although individual storms are definitely short-lived weather phenomena, climate patterns affect storm formation, intensity, and frequency. The climate impact of tropical cyclones, and how changes to climate will impact cyclones, is an area of active research.[13] Tropical cyclones exist by a coupling between atmosphere and ocean. Cyclones get their energy from the ocean surface, through evaporation of moisture and release of the latent energy in deep cloud systems. The vertical motion in these clouds organizes into spiral bands of clouds and a large-scale flow through and out the top of the storm. This is hard to represent without the correct coupling between the surface and the ocean. Tropical cyclones are very destructive and disruptive to societies where they make landfall. They may also be important for large-scale transport of moisture both in the vertical to the upper troposphere and in the horizontal to higher latitudes.

---

[13]For a recent summary, see Knutson, T. R., et al. (2010). "Tropical Cyclones and Climate Change." *Nature Geoscience, 3*(3): 157–163.

Tropical cyclones are barely resolved in global climate models at high resolution (25 km, 16 miles), and usually without the correct intensity (they are too weak). Some of the efforts to make variable resolution models are driven by a desire to have high resolution in tropical cyclone active basins.

8.4.2 Monsoons

The seasonal **monsoon** circulations are giant sea breezes that provide seasonal mois-ture to many regions of the planet.[14] The South Asian or Indian Monsoon is one of the most important: providing a lot of the seasonal rainfall that feeds a billion people. The South Asian summer monsoon is a consequence of heating of the land mass of South Asia (the Indian subcontinent), causing onshore flow from the surrounding oceans. The convergence and uplift from mountains cause significant precipitation. This releases latent heat, causing more upward motion, and thus feeds on itself. The flow over the ocean is strong enough to affect ocean currents. The topography of the Indian sub-continent, the Himalayan plateau, and even the East African highlands contribute to the development and evolution of the monsoon. The process is a seasonal combination of the atmosphere, ocean, and land surface in the tropics, driven by seasonal radiative processes. Monsoons exist in other regions as well (Africa and even southwestern North America), with similar processes, but smaller magnitude. The monsoons are seasonal and occur every year. They are affected by longer-term climate patterns as well. Monsoons are critical, because most monsoon regions have societies that have grown to depend on them, and failures of the monsoon rains can be disastrous.

Monsoons are affected by ocean currents and by topography. Both of these are difficult to represent in the coupled climate system at large scales, and many global models struggle with the details of the South Asian Monsoon. In particular, the biases in ocean circulation and lack of resolution of mountains can contribute to different strengths of the monsoon, different convergence over India, and hence very different rainfall.

8.4.3 El Niño

One of the best known interannual patterns of climate variation is the **El Niño Southern Oscillation (ENSO),** named for the warm waters that occur off the coast of South America every other December or so (around Christmas: *El Niño* means "the boy" in Spanish, referring to the Christ child).[15] The warm water is a conse-quence of a coupling between the ocean and atmosphere in the tropical Pacific.

---

[14]The World Climate Research Program (WCRP) has a good factsheet on monsoons; see http://www.wcrp-climate.org/documents/monsoon_factsheet.pdf.

[15]A good El Niño overview with current state, forecasts and factsheets on what El Nino is available from the U.S. National Oceanic and Atmospheric Administration, http://www.elnino.noaa.gov.

Normally, winds blowing westward along the equator push warm water into the western Pacific, and causing cold upwelling of deep ocean water near South America (see Chap. 6, Fig. 6.2). The atmosphere responds with rising motion over the warm water in the west, with formation of clouds and rain, while air descends in the east.

During an El Niño event, the westward wind is disrupted and the warm water flows east to South America. The pattern of rainfall moves toward the central Pacific. The mechanisms for this are complex; they have to do with a combination of winds in the atmosphere and slow motions of the mixed layer in the ocean. When too much water piles up in the west, and the thermocline (the bottom of the mixed layer) gets too "tilted" from west to east, internal ocean waves can result. The waves are also affected by wind patterns. The tropical Pacific is a giant bathtub that sloshes around with wind blowing intermittently over the top: When the wind hits the sloshing wave just right, it can amplify and reinforce the wave. The opposite phase, with cold water near South America and warmer water in the western Pacific, has been termed *La Niña* ("the girl" in Spanish). La Niña brings more rain to the western Pacific.

Representing this coupling between atmosphere and ocean has been a difficult task for coupled models. To get the period right, slow and large-scale wave motions in the ocean need to be simulated, and their effects on the atmosphere represented.

8.4.4 Precipitation and the Land Surface

Another significant climate coupling involves relationships between rainfall and surface conditions. Precipitation is coupled to evaporation: Water has to come from somewhere to get into the atmosphere. In coastal regions, this is from the ocean, but in continental regions far from the oceans, water is recycled. The land surface brings water back to the atmosphere through evaporation and transpiration from plants. Precipitation is most tightly coupled to the surface in semiarid regions, where there is enough energy to evaporate water and enough water in the soil to be released by plants (transpiration) and evaporation. Drought can result from this system's being out of balance: Less rainfall can dry the soil and damage the ability of plants to return the moisture to the atmosphere, creating a cycle that may lead to drought. Many regions of the world are prone to such couplings and multiyear droughts, including the Sahel of Africa (south of the Sahara) and southwestern North America.

8.4.5 Carbon Cycle and Climate

On timescales longer than a few years, there are significant couplings between the carbon cycle and climate.[16] The carbon cycle governs the sources and sinks of

---

[16]For more on carbon-climate coupling, see Archer, D. (2010). *The Global Carbon Cycle.* Princeton, NJ: Princeton University Press.

atmospheric CO_2, which is a greenhouse gas. On the scale of decades to centuries, increasing CO_2 can cause enhanced plant growth, which may damp increases in CO_2 by fixing more of it in the terrestrial biosphere. Enhanced plant growth is dependent on water and nutrient availability, so the coupling is also dependent on the hydrologic cycle and other biogeochemical cycles.

On timescales of thousands of years, carbon in the ocean adjusts to the ocean circulation. It is thought that the carbon cycle amplified the forcing from the sun that ended the last ice age by a change in ocean circulation. A perturbed ocean circulation near the end of the last ice age resulted in more carbon being released to the atmosphere. Simulating these effects is possible in climate system models, but it requires long simulations. Understanding these couplings of the past carbon cycle is critical for testing climate models and enhancing confidence in future predictions.

8.5 Challenges

There remain many challenges in coupling together the different components of the earth system. The difficulties with coupling components into a system that accurately represents the earth's climate system are really a combination of uncertainty in process representation, and uncertainty from observations (How do we know when the system is "right"?). The strategy has generally been to develop and test component models (e.g., atmosphere, sea ice, terrestrial system) with "observations" until good solutions are achieved, and then to couple them together. If the models are not able to reproduce observations (of their respective spheres) with the right boundary conditions, coupling will be hard. But even if they do reproduce observations, small inconsistencies in the observations can result in systematic errors in the coupled system.

A variety of different complexities and challenges have been detailed in the coupled system. Some of the hardest complexities are those processes and features that extend across different components, and where the interactions between different components are critical. Some of these interactions are at the small scale, such as feedbacks between precipitation and the land surface. Some are on short timescales, such as coupling between the surface ocean and the atmosphere that helps cause tropical cyclones.

Getting these coupled processes correct is an important prerequisite for understanding and simulating how such processes will change. Getting present coupled processes correct is not a guarantee of predicting their future. It is sometimes called a necessary but not sufficient condition. Sometimes models show projected changes as differences from the present day, such as the change in surface temperature in a region. But if the surface temperature has a systematic error in the present in the model, especially if the error is larger than the projected change, then caution is warranted.

Some of the processes have medium timescales: seasonal, like the monsoon, or every few years, like El Niño. El Niño is a coupled atmosphere-ocean mode with a

3–5 year timescale. And some of these processes act on climate change scales of centuries, such as sea-level changes and ocean feedbacks. Finally, some processes like the carbon cycle have important interactions on geologic timescales.

Naturally the development of coupled models has shifted from asking shorter-term to now longer-term questions. Basic issues of interactions and coupling to maintain a stable climate have been achieved in most models, and they are now trying to simulate coupled modes of variability and to understand the longer-term evolution of the climate system on the century scale. This is the scale of climate change over a human lifetime or several generations, and it is where climate models are being used, and where their uncertainty is being assessed. Of course, these predictions rely on representing coupling processes correctly (such as surface exchanges between the ocean or land and atmosphere).

Of these predictions, sea-level rise and carbon cycle changes present some of the biggest challenges. Many of the processes for simulating ice sheets are not well understood (see Chap. 7), and these models are still fairly new. This means a great deal of uncertainty exists, and this is an area where projections are still changing. In addition, the feedbacks between climate and the carbon cycle on century timescales are uncertain: In principle, plants will take up more CO_2, reducing any increase in the atmospheric concentration that creates a radiative forcing. But because in practice plant growth may be limited by the availability of water and nutrients, plants may not take up more CO_2. Uncertainties in future sea level, and uncertainties in the uptake of CO_2 in the carbon cycle, are dominated by specific processes. This is actually a good thing. Attention is focused on understanding and simulating specific processes in the climate system. This is one way climate models can be used to explore specific predictions and help improve understanding.

8.6 Applications: Integrated Assessment of Water Resources

This case study explores coupling models of the earth's climate with models of human activities. This type of coupling takes place at many spatial and temporal scales with many coupling strategies. The coupling of climate models with models from other disciplines is often called integrated assessment modeling[17] and integrated environmental modeling.[18]

Integrated assessment models provide information that helps guide decision making; they are also used to investigate the consequences of decisions. The process brings together many disciplines. Integrated assessment modeling increases

---

[17]The Royal Society, London, "Modeling Earth's Future," October 1, 2013, https://royalsociety.org/policy/publications/2013/modeling-earths-future/.

[18]The International Environmental Modelling and Software Society, http://www.iemss.org/society/index.php/scope.

the complexity of the modeling environment by adding different disciplines. This can be thought of as adding more detail to the human sphere, or adding other spheres to a coupled model. In the process, the representation of the human-natural system is simplified relative to physical models or economic models on their own. Not only is the range of disciplines increased when human systems are simulated, but uncertainty sources become larger and are influenced by intentional and unintentional decisions of humans.

Human system models are much less constrained than climate models of the physical earth system. Modeling assumptions do not have the definitive cause-and-effect relationship of physical principles: Money is not conserved like energy and mass, and the range of possible outcomes is constrained not by physical laws, but by economic principles. Thus, the ability to apply integrated assessment models directly in deterministic decision making is even more difficult than the application of physical climate models. There is argument about just how useful global integrated assessment models focused on energy, economics, and agricultural are for decision making.

Perhaps the best way to use an integrated assessment model is not to generate a specific answer to a policy question, but to provide insight about crucial interactions and uncertainties between human and natural systems.[19] In other words, integrated assessment models highlight (a) specific climate impacts that might drive societal responses, (b) what aspects of society are affected by different climate impacts, and (c) the relative sensitivity of human systems to different factors. Described later in this section is a specific example of water resource management.

Integrated assessment and integrated assessment modeling is broader than the coupling of climate models with macroeconomic models mentioned in the chapter body. The National Research Council defines integrated assessment as "a collective, deliberative process by which experts review, analyze, and synthesize scientific knowledge in response to user's information needs relevant to key questions, uncertainties or decisions.[20]" Integrated assessment is a problem-solving methodology to bring together natural science, social science, and policy to support knowledge-based decision making.[21] It is a form of participatory, iterative problem solving, as discussed in Chap. 12. From the perspective of the climate scientist, integrated assessment is a structured process that inserts knowledge and consideration of climate change into decisions such as building and maintaining infrastructure; forest management; and water resources for agricultural, industrial, and human consumption. Climate change is often placed into the context of existing policy, built infrastructure, and known weather vulnerabilities; hence, climate

---

[19]Morgan, G. quoted in "Modeling Earth's Future," Royal Society, London, 2013, https://royalsociety.org/policy/publications/2013/modeling-earths-future/, p. 22.

[20]NRC. *Analysis of Global Change Assessments—Lessons Learned.* National Academies Press, 2007.

[21]Graham Sustainability Institute, "Integrated Assessment," http://graham.umich.edu/knowledge/ia.

change provides an incremental alteration of an existing end-to-end system already in place.

Integrated assessment is frequently applied to water resource management and has been used widely to consider changes in vulnerability to water resources in the western United States. An example is the *California Integrated Assessment of Watershed Health*, sponsored by the Environmental Protection Agency.[22] This study considered four vulnerability factors associated with watersheds: climate change, land cover change, water use change, and fire vulnerability. With regard to climate change, vulnerability indicators are developed for precipitation, mean temperature, minimum temperature, maximum temperature, snowpack, minimum flow (or baseflow), and surface runoff. These indicators show the requirements of the climate model to provide not only direct measures of climate change (temperature and precipitation) but also derived integrated parameters (snowpack, baseflow, and surface runoff).

Climate-change vulnerability is mapped spatially, from low to high, on a 2050 time frame and then considered in context with other vulnerabilities noted above. Climate-change vulnerability is highest in the northern third of California, where temperature increases cause large alterations to snowpack, minimum flow, and surface runoff. A conclusion from this work is that preventing landscape degradation in relatively unpopulated areas at the headwaters of rivers increases the resilience of both ecosystems and human systems to climate change.

8.7 Summary

The climate system can be simulated in many ways with different types of models. Some of these models are dynamical: what we commonly think of as climate models, models that look and work like weather forecast models. There are also statistical models of climate. These are often used for downscaling projections of a large-scale model to smaller scales. Downscaling can also be dynamical. Regional climate models are examples of smaller-scale dynamical models used to generate more detailed statistics, and run with boundary conditions from larger-scale models. Coupling components of the climate system has evolved over the past 40 years or so. First, separate component models are run with fixed other parts of the system, such as running an atmosphere with fixed sea surface temperatures. Coupling is the process of trying to tie component models together. Errors are now small enough that such coupling does not generally cause the climate system to drift, but there are still uncertainties in observations that go into evaluation of the coupled system.

---

[22]U.S. Environmental Protection Agency. *California Integrated Assessment of Wathershed Health*, November 2013, http://www.mywaterquality.ca.gov/monitoring_council/healthy_streams/docs/ca_hw_report_111213.pdf.

The climate system is full of interactions between the different components. Coupled systems collect and pass the information around, and they ensure energy and mass conservation. These interactions are manifest especially in the hydrologic cycle and the carbon cycle.

On the one hand, these complex coupling mechanisms make it difficult to simulate the earth system. On the other hand, representing these phenomena correctly can be a strict test of climate models. Proper simulation of the carbon cycle, tropical cyclones, and the right period and amplitude of El Niño events are all strong indications that climate models can represent various modes, timescales, and important processes in the earth system. In Part III, we further examine uncertainty in the models and ask how good they are at these various processes, and why should we trust them for the future.

Key Points

- Climate models can be global or regional.
- In addition to dynamical system models, statistical models of climate can be used.
- There are complex interactions in the climate system, including for water.
- Coupled effects such as sea-level rise are difficult to simulate.
- There are many timescales of interactions between component models.

Part III
Using Models

Chapter 9
Model Evaluation

The preceding chapters gave an overview of the climate system and its components as well as a primer on how we simulate those components, but even so we've just skimmed the surface. Why should we trust a climate model? Generally, we gain trust in a model through evaluation and validation of the model. We then use the model to make projections of the future. In this chapter, we describe the basics of how climate models are evaluated and how they are tested. The language and terms used in this discussion can be confusing. For example, the terms *validation* and *evaluation* are often used to mean different things, and a *projection* is not the same as a *forecast*. We will see why shortly. Testing models is a critical part of the development process.

9.1 Evaluation Versus Validation

Evaluation is the process of understanding a model and how well it works for a specific purpose. It is the process of ascertaining the value of a model. Since a model (whether a blueprint or a physical model of an object like a car or an engine) is a *representation* of an object, it is usually not an exact replica in some way. In other words, the model contains some simplifications. **Validation** is the process of ascertaining or testing the "truth" of a model. And since all models are incomplete representations of reality, we are not really seeking a perfect representation of the truth. Instead, we are seeking the value to be found in an imperfect representation provided by a model. Often the key aspect of value comes from knowing how good or bad the model is relative to observations. The goal is really to figure out what *value* a model has (by evaluation). The value depends on the application, as we make clear in this and later chapters.

Look at common models in the world around us. A picture, even a photo, is an imperfect representation of a three-dimensional object, but our brains use pictures as a model to understand objects. The picture or model is not the actual object. It might be a scale model of a building, or a schematic picture of two parts of a piece of furniture that will fit together. Thus, all models are incomplete or wrong in some way. For example, you can't sit in a scale model of a chair, and the strength-to-weight ratio of a scale model is probably very different from that of the

© The Author(s) 2016
A. Gettelman and R.B. Rood, *Demystifying Climate Models*,
Earth Systems Data and Models 2, DOI 10.1007/978-3-662-48959-8_9

actual chair. But most models are useful, even if they are wrong in some way. They tell us something about the object or system being represented. In other words, a model of a structure is built so that we can better understand what something will look like. Schematic diagrams help us understand how to put two pieces together. We evaluate models to understand how well they represent particular aspects of the system: *These representations have value.* The better the model for a particular metric, the greater its value for a particular purpose.

9.1.1   Evaluation and Missing Information

The evaluation process is usually indirect, and there is often uncertainty in what we are evaluating the model against. We have to evaluate a climate model against the climate system using imperfect and incomplete or missing information. Generally, we do not have a complete and accurate description of the climate system. Because climate is a statistical measure (the distribution), we have to build up statistics. Our statistics may not be complete, especially for extreme events. For instance, what is a 1 % chance of having a certain amount of rain (a lot or a little) in a season? If we only have 50 years of records, we do not really know: The lowest or highest seasonal rainfall is a 2 % chance if it is random (1/50).

We generally have distributions that are not well described. We may be missing critical information. For climate, we need information in the past, and we cannot go back and collect more information. If we are concerned with the climate somewhere, but we have no records, it is hard to describe the distribution of climate. This is incomplete information that cannot be taken again.

In addition to the lack of information, the observations we do have are generally not perfect: Observations contain errors. If the errors in observations are known, they can be corrected for. A great deal of work is done to test and evaluate observations to understand errors and ensure accuracy. Where it can be difficult is if the errors are unknown, and if the errors (particularly unknown errors) change over time.

A good example of observational error is the measurement of temperature. Of course records can simply be missing: There were no records of temperature before a practical thermometer was invented and used in the Middle Ages.[1] Galileo Galilei was one of the first, in the late 16th or early 17th century, to develop a liquid-filled tube that changed volume with temperature. But it took another 100 years or so to agree on a standard unit of measurement. Daniel Fahrenheit (1724)[2] and Anders Celsius (1742)[3] both proposed scales that are still in use (with some variations)

---

[1] For a description of the development of the thermometer, see McGee, T. D. (1988). *Principles and Methods of Temperature Measurement.* New York: Wiley Interscience.

[2] If your Latin is good, see Fahrenheit, D. G. (1724). "Experimenta et Observationes de Congelatione aquae in vacuo factae." *Philosophical Transactions of the Royal Society, 33*(381–391): 78. doi:10.1098/rstl.1724.0016.

[3] Described in Benedict, R. P. (1984). *Fundamentals of Temperature, Pressure, and Flow Measurements,* 3rd ed. New York: Wiley.

today. But thermometers have changed over time, and although temperature records exist back to 1800 or so, their accuracy is very different from thermometers in use today. Even modern instruments are different from each other and change over time (discussed later in this chapter). How does this affect climate records? Suppose a temperature measurement is conducted over many years at the same place. Suppose there is a systematic error, maybe the liquid in the thermometer doesn't rise or expand as expected, and it reads a colder temperature than the true temperature. If a new and more accurate thermometer replaces it, then the more recent records will record a higher temperature, and we might conclude the temperature has warmed. But the error in the measurement has just changed.

To evaluate a climate model, we simulate the climate of a particular point in the system with a model and then compare that simulation against a set of observations. Recall that *climate* is the distribution of something like temperature or precipitation at a particular point. Thus, models are evaluated against the distribution of possible states, not just the mean state. Often what we really care about is how the model simulates extreme events such as floods, heat waves, droughts, or tropical cyclones. We may not care about the mean.

We construct a distribution of temperature or precipitation observations to compare our model against. But those observations may have either systematic errors (like a bias in the observation, shifting the mean of the distribution), or there may be random errors in the observed distribution (see Chap. 1). The observed distribution may be undersampled, particularly for extreme events (see Fig. 9.1). Figure 9.1a is a sparse distribution. There are a small number of points (25) drawn randomly from a distribution, where the probability (vertical axis) of a value (horizontal axis) is what we want to find out. The distribution has a most likely value at 100. Figure 9.1b shows the probability distribution function from the sparse sample: There are no points greater than 104 or less than 97. Figure 9.1c shows 3000 samples from the same distribution, whose probability distribution function is represented in Fig. 9.1d. Now it is clear the most likely value is 100, and a small but significant number of points have values less than 97 or greater than 104. The problem represented in this figure is similar to the question: If we are trying to represent the extremes of a distribution that occur infrequently (once every 100 or 500 years, for example, like for a "500-year flood"), how do we know what those probabilities are from 50 years of data? This is a problem particularly for understanding and evaluating extreme events.

What does it mean to evaluate a model for prediction? If the model is wrong in some way, you need to know that. The key question is whether the model is accurate for what you want to predict. If you want to predict the climate in the tropics, you might not need a detailed model of sea ice or of snow cover. If you are in polar regions, it is critical to have a detailed representation of snow and how it absorbs and reflects energy from the sun. But the climate system is interconnected, so there are limits to what can be ignored, and climate models seek to represent consistently the entire possible set of states of the system. Here's another example. Say that you want to predict the weather for a few days. You can probably fix the carbon dioxide concentration in the model. You may also not need to worry about

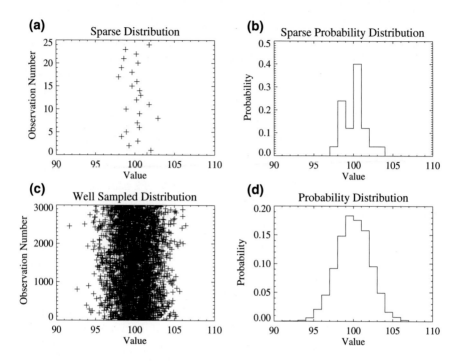

Fig. 9.1 Sampled distributions. Points representing individual observations are randomly sampled from a distribution with a mean of 100 and a standard deviation of 2. **a** Sample with 25 points. **b** The probability distribution function (PDF) of these points. **c** Same distribution with a sample of 3000 points. **d** The PDF of these points

small errors in the energy budget in such a model. We need to evaluate models for a purpose and assess whether they are useful for a particular purpose. The weather prediction model that does not conserve energy may be fine for 48-h forecasts, but it is likely not a great climate model.

Evaluation of models also involves comparison of different models. There are about 25 different climate models of varying complexity that help inform our understanding of global climate. In Chap. 11, we discuss details of how these models are related or independent, but they represent the best estimates of the climate system. Each estimate will be different since the representation of the system is quite different. We can also evaluate models against each other.

9.1.2   Observations

For models, the primary evaluation method is to evaluate the processes or results of models against observations. **Observational uncertainty** is a key problem. Observations are biased due to **sampling uncertainty** (gaps in records), as illustrated in Fig. 9.1, but also due to systematic errors in measurement. It is often as important to know the uncertainty of a measurement as it is to know the numeric value of the measurement. Sometimes, knowing the uncertainty is even more valuable: If you do not understand the uncertainty in an observation, it is not possible to understand if a model is statistically the same (correct) or different (wrong) compared to an observation. If the mean temperature is 68 °F (20 °C), and a model predicts 72 °F (22 °C), is the model wrong? If the expected error or uncertainty in the observation is ±4 °F (2 °C) or larger, the model is correct. If the observed uncertainty is smaller than ±4 °F (2 °C), then the model is wrong.

Figure 9.1 addressed the sampling uncertainty of not knowing the "true" distribution. Figure 9.2 illustrates the difference in distributions. If there is a lot of variability or spread about the mean (which can be measured statistically by the **standard deviation**; see Chap. 10) in the model and observations, then the model is not statistically different from the observations at some level of probability. Separating the black and blue curves is hard; separating the red and blue curves is easier, even though the red and black curves have the same mean.

Scientists often try to estimate a confidence level, or confidence interval, for a distribution as a way of understanding the expected error. If an observation has uncertainty, a 95 % confidence interval indicates we are 95 % certain to be within a given range. In Fig. 9.1d, this range is about 95–105, so 5 % of the observations fall outside this range. When comparing models to observations, if the confidence interval for the model overlaps the observation, then the model is not significantly

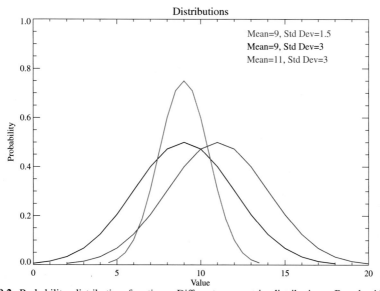

Fig. 9.2 Probability distribution functions. Different symmetric distributions: Broad with the same width (*black* and *blue*) but different means and sharply peaked (*red*) but with a different standard deviation. *Red* mean of 9 and standard deviation of 1.5. *Black*, mean of 9 and standard deviation of 3, *Blue* mean of 11 and standard deviation of 3

different (statistically) from the observation. Often the model statistics are better known (sampled) than observation statistics (the observations are sparse). For example, if we have 25 years of observations, we get the samples in Fig. 9.1a. But with a model, we can run it for 3,000 years to get the samples in Fig. 9.1c.

Models are evaluated not just on the mean state, but also on their representation of variability. Refer to the distribution functions in Fig. 9.2. You can match a distribution function mean value (the black curve) with another distribution with the same mean, but very different width or variance (the red curve). Even the shape can be different (symmetric or skewed). The mean may be the same for the red and black curves in Fig. 9.2, but the black curve has twice the spread of the red curve (higher variance). Thus, the black curve has a higher probability of extreme events than the narrow red distribution. Even if the means are similar, if the distribution represents temperature, for example, the climate is a lot different.

The key to evaluating models is to collect observations and the uncertainty in the observations, and then compare the model as closely as possible to the observations. Getting the different statistics (mean, variability) correct is critical. Which statistics are important will depend on the application.

It is also critical to compare like elements of a model with their corresponding observations. One should not compare apples with oranges, meaning like for like comparisons are critical. What does this mean in the context of climate observations?

Often observations are not what we think they are. Most observations contain a model themselves. Consider the following examples, all of which are trying to measure the same thing: the surface temperature of the ocean.

1. A liquid thermometer contains a substance like alcohol or mercury that expands and contracts with temperature. This thermometer measures expansion (volume). To convert this change into a reading, the thermometer needs a scale: a model for how the material should expand or contract with temperature. Put alcohol (often colored red) in a mercury thermometer (silver liquid), and you get the wrong answer.
2. An electronic thermometer contains a piece of metal (called a thermocouple) that has different electrical properties (usually resistance) with temperature. This thermometer measures electrical resistance. The device has a "model" of how a different resistance corresponds to temperature. The model has uncertainty in it. How much? That may or may not be known.
3. A satellite orbiting the earth sees the emission of the sea surface at a given wavelength of light as an electronic current generated when a number of photons (light particles) of a given wavelength (in the infrared, in this case) hit the detector (usually like a fancy digital camera). The detector converts photons into electrical charge (current). This thermometer measures electrical current. The number of photons received is a function of the surface temperature of the ocean over a given volume, but also of the atmosphere above it. A model is built to understand the temperature. It is supposed to correspond to the electronic or liquid thermometer stuck into the ocean at the same point, or at all points over the region where the satellite is sensing.

4. You can also build a thermometer like the satellite that senses energy in a distinct wavelength to read the temperature (an infrared thermometer). This device may perform more like the satellite, but it is seeing only a few square feet (1 square meter) of ocean, whereas a satellite may be collecting photons from many square miles. This thermometer also measures electrical current.

All of these measurements contain models that translate a measurement (volume, resistance, current) into temperature. These observation models are different, and they may contain errors. These errors may be due to the imprecise nature of measuring volume for example, or they may result from the distribution of temperatures in the field of view of the satellite.

Each observation is measuring a different mass of water. For example, the thermometer is measuring the small region of water around the "bulb." The thermocouple has a similar sampling area, but maybe a different response time to reach a constant temperature, and maybe it is stuck deeper or shallower in water. The infrared thermometer may measure a region that is 1–20 ft (0.3–6 m) across on the surface rather than, say, 1–3 in. below it, and the satellite is measuring the surface emission of maybe 0.6–30 miles (1–50 km) of the ocean. Even if all of these temperatures are correct, they measure the thermal energy of different water molecules.

So what temperature does a climate model use for the ocean surface? The model might have a temperature of the top layer of the ocean, but that layer might be 30 ft (10 m) thick. A satellite or infrared thermometer at the surface sees the emission from just a small thickness of the surface. As anyone who has been in a stratified lake or ocean knows, the average temperature of a thick layer of water below the surface may be much colder than the temperature at the top. The model is representing the heat content of the entire layer. Often models have a "skin" temperature to more closely match observations. Another way to do the comparison is to estimate not the temperature from the model, but rather the thermal emission of photons that would result from that temperature, and this can be compared directly to the satellite values before a temperature conversion.

The direct simulation of an observation (simulated numbers of photons, as in the example above) is a particularly useful means of comparison when comparing complex observations. Consider the properties of clouds. The model that the satellite uses to turn photons in a wavelength into a description of a cloud (like the amount of liquid water) can be used in reverse to take the climate model cloud and determine how many photons at a wavelength it should emit. This makes the comparison more robust (like comparing red apples to green apples).

The process of evaluating a model, and especially a climate model, thus has several steps. First is to collect observations and then to analyze and understand the observations. Understanding includes accuracy of the observations, and the uncertainty in the observations, which comes from the length of the observational record (the sampling in time) and the sampling in space as well. There are likely to be many different observations to compare with a model. Some observations may be the same quantity, such as several sets of temperature observations. Some observations will be different quantities, such as temperature, precipitation, wind, or soil moisture. Once we have a set of comparisons with observations, we know how different a model is from observations in many ways. We can decide the value of the model and whether the model is adequate by having a value higher than some threshold, such as a sufficiently low error in comparison to an observation. And if not, or if we are not satisfied, then we seek to improve the model. In a theoretical sense, since all models are wrong in some way, they can always be improved. Of course, the same could be said for observations. Understanding the uncertainty in the observations, including the models that go into the observations, and the difference between what the observations and the model represents is critical for evaluation.

9.1.3   Model Improvement

Specifically for climate models, there are many different ways to approach model improvement. The methods come from an understanding of where the model error comes from. Models are a series of components (atmosphere, ocean, land) coupled together. Each component model is a series of processes (e.g., radiative heating, motions, transformations of water, plant growth) described by parameterizations of processes (condensation, evapotranspiration, etc.). The processes themselves may not be represented well. Perhaps the major issue is not describing the basic physics or chemistry of a problem, but rather the variability below the model scale. An example might be a chemical reaction with a defined reaction rate. These rates are measured usually in a laboratory with pure substances. But approximating the same rate in a large volume of atmosphere, which is not well mixed, may be very difficult to get right. Thus, parameterizations of processes can be improved, often with detailed observations of the world around us to see if we can reproduce particular times and places. Comparisons can be made for individual events, or for many events to generate an "observed" and "simulated" climate (or climatology) of a particular place or many places.

Model errors (model uncertainty) can also arise from the complex coupling of the system rather than from individual processes. Individually the atmosphere can be driven by surface observations, or it can be coupled to the land and ocean. Often there are errors in the coupling or the translation that can lead to biases. More frequently, there are biases in one component or process that affect others, and often *compensating errors* arise where one process may be too large, and a competing process too small, with the result being right, but for the wrong reasons. For example, if there are too many clouds shading an ice-covered surface, but the surface is darker than it should be, then you can get the right surface energy balance, but for the wrong reason. The model appears fine compared to observations. But how this incorrect balance changes may be different. The hope is that with sufficient observations (e.g., also measuring the brightness or albedo of the surface), these biases can be eliminated.

When evaluating models, we typically find that they differ from each other, as well as from the observations. This is actually useful. Different climate models contain different representations of different processes, coupled together in different ways. It is not surprising that answers will be statistically different when compared. This applies not just to large-scale climate predictions like the global average mean temperature well into the future, but also to evaluation of individual events. The differences are useful, in the same way that the different models of temperature from a mercury thermometer or an infrared thermometer may be useful for understanding the uncertainty and variability of temperature.

9.2 Climate Model Evaluation

It is not always straightforward to envision how a climate model is evaluated. The concept of evaluation involves comparing a climate model to observations. Climate is a distribution, so the process of evaluation is a comparison of distributions, for example, a distribution of temperature, between a model and a set of observations. We have discussed how observations are taken, but how are climate model data generated to create a distribution? There are different ways to perform a simulation experiment that integrates a climate model. Since climate models are computer programs, performing a simulation is usually called *running* the model (as in running a computer program).

Climate models are generally run in different ways for evaluation against observations in the present and past, than for prediction of the future (for which there are no observations). We discuss some of these ways in the sections that follow. The different types of simulations are designed to test different parts of climate models against different types of observations. Understanding the ways that simulations are run is important for understanding and evaluating model output, and for understanding the results.

9.2.1 Types of Comparisons

There are several different types of simulations for evaluation of climate models. Typically parts of a climate model (like the atmosphere) are constrained in some way (some inputs are specified, such as the ocean surface temperature beneath the model) to evaluate them against some type of observations for the present or the past. These can be observations of individual events or case studies. They can also be representations of climate (averages, variability) over short or long periods of time, from years to centuries.

Representing individual events is one way to test models. We can try to estimate individual events such as a particular tropical cyclone (i.e., Hurricane Katrina or Hurricane Sandy) by starting up the models with observations from just before an event, and comparing how the model does. Weather forecast models are typically evaluated routinely in this way: How well did they do in "predicting" the weather 1 day (or 2 or 10 days) in advance? Such **hindcasts** are also valuable for improving parameterizations and representations of processes in a model, maybe changing how the surface exchange between the atmosphere and ocean makes a better representation (i.e., better surface wind speed or precipitation) of a particular storm. Maybe this applies to all tropical cyclones. Weather models typically undergo rigorous testing in this manner to generate error statistics and improved forecasts. For some of the key aspects of climate models, we do not have complete verification of our forecasts. We cannot easily evaluate feedbacks due to climate forcing, for example. The verification piece is hard and has to be approximated.

We can also apply the same comparison to multiple models. For example, what is the range of model simulations of a particular event or storm, or all storms? Evaluation is typically conducted for a purpose, since we cannot evaluate everything. If you want to know how tropical cyclones might change in response to forced climate change, then evaluating the representation of cyclones in current models is quite important.

These types of experiments can be performed as well with other components of an earth system model. Observed temperatures and precipitation can be fed to a land surface model to try to reproduce observations of soil moisture. Or an ocean model can use specified air temperature and winds to try to reproduce ocean currents.

All of these comparisons can be done for long or short periods of time. For long periods of time (20 years of observations, for example), the models are used to generate climate statistics (probability distribution functions) that can be compared to observed distributions. But models (even climate models) can also be evaluated using short-term forecasts to try to predict the details of weather events within the timescale of a few days (just as weather models are evaluated). It is often easier to focus on particular well-observed locations and evaluate specific cases, or a set of cases. In these evaluations, models are run like weather forecast models for a few days, and the statistics of the agreement are evaluated. Using many forecasts (starting every 6 h) and running for only a few days can be fruitful. Many of the errors in processes in climate models (like clouds) show up in just a few simulated days. So different parameterizations can be rapidly tested using short simulations.

9.2.2   Model Simulations

We have described different ways to run climate models for evaluations. These usually involve constraining the climate model in some way to better represent the observations. Or one component is replaced with observations (e.g., fixed sea surface temperatures), and the rest of the model is "forced" to use observations as boundary conditions. Ultimately, none of the components is fixed and the model runs with all its components coupled together. But there is still forcing of the system, usually provided by parts the model doesn't simulate, such as specifying greenhouse gases like carbon dioxide and methane, or estimates of the emissions of sulfur gas (sulfur dioxide) by volcanoes, or even small variations in the solar energy that reaches the earth. In this way, model simulations can be designed to try to reproduce the past of the entire climate system.

Reproducing the past can mean reproducing the last 100 years or so when we have some observations or even the last 5 years. It can also mean reproducing the past a long time ago: before measurements were available directly. The oldest records from thermometers go back to about 1750 or so in a few locations. Before this, only proxy records are available. **Proxy records** are indirect records of a process sensitive to a climate variable (usually temperature or precipitation). For example, the width of tree rings is often proportional to seasonal temperature and/or precipitation at a location. Assuming the trees respond the same way in the past as they do now, records from tree cores can provide seasonal climate information going back hundreds or even a few thousand years. Present climate can be related to tree ring growth rates over the last 100 years. The relationship (as a statistical model) can then be used to estimate temperature and precipitation records at locations in the past where there are tree core records, but before instrumental records. The longest lived trees go back nearly 5,000 years (bristlecone pines) but such proxies are more common for several hundred years (a Ponderosa pine tree can live 600 years).

Figure 9.3 shows a series of different proxies related to the instrumental record that goes back to 1750. Boreholes are deep holes that measure the temperature down for several hundred feet (100–200 m). Since heat moves slowly in the ground, these borehole temperatures can measure variations in local temperature back 500 years or so. Historical records come from human histories about drought, famine, volcanoes, or other historical events that might impact climate. For example: the eruption of Mount Vesuvius in 79 A.D. Ice core records go back

Fig. 9.3 Paleoclimate proxy records. Different types of proxies are shown with the time space (in years before present) covered by the proxy on the horizontal axis. The time scale is logarithmic. Data from the National Oceanic and Atmospheric Administration Paleoclimate Program

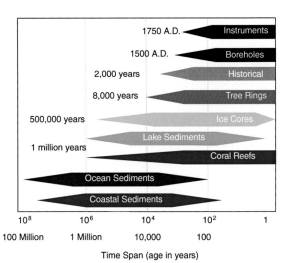

500,000–800,000 years. Coral reefs and the remains of coral go back nearly a million years, as do sediment records in lakes, oceans, and coastal regions. These sediments contain different species of microorganisms that survive well in different temperature conditions.

For proxy records like tree rings, there is always a model that translates the proxy into a climate variable, such as how the width of tree rings translates into precipitation and temperature. Proxies are calibrated on present-day conditions but then extrapolated to different conditions in the past or future. Note the dangerous word *extrapolate*. The assumption is that the past behaves like the future.

Ice cores can also provide some estimates of temperatures, because the isotopic composition of the water molecules in the ice is related to the formation temperature (see Chap. 3). Heavier isotopes (atoms with extra neutrons: hydrogen with a proton and neutron, ^{2}H, instead of just a proton, ^{1}H or just H) move differently between liquid and ice phases as a function of temperature. Ice cores also provide past records of stable gases in the atmosphere like carbon dioxide and methane in gas bubbles trapped in ice. These records go back up to 800,000 years (see Chap. 3, Fig. 3.2). Records of sediments (such as pollens) can go back even farther. These records of **paleoclimate** (*paleo*—comes from the Greek word for "ancient") are not direct measurements of climate, but are proxy records related to climate. So comparing them to models set up to run for the past is instructive but subject to the apples-and-oranges problems discussed earlier, and usually take a fairly complex model (often a statistical model) to interpret. But it is useful for climate models to find other climates in the past to simulate, as a way of evaluating models for their representation of the future.

9.2.3   Using Model Evaluation to Guide Further Observations

An important aspect of testing observations is knowing what to measure and where to evaluate a model. Models can help us understand what to measure. By looking at where a model is most sensitive—that is, where small changes in the model itself or the initial conditions result in big changes in results—we can find the places and conditions where we need observations to be able to evaluate (and constrain) a model. In weather forecasting, for example, there are certain situations where we know that small uncertainties can lead to big errors. A classic example is for tropical cyclones: Small differences in the temperature and pressure field around a storm govern how it will intensify or weaken, and in what direction it will move. To improve forecasts, aircraft fly around the storms and take additional observations. These aircraft are guided by forecast models that identify where additional information can make the most difference to the accuracy of a forecast.

In a climate context, we often do similar things on a larger scale: What climate phenomena do we not understand? And what observations would better constrain the phenomena? We know, for example, that clouds are an important uncertainty in models. We also know they are poorly sampled. This guides field projects into critical cloud regions for climate (such as low clouds over the ocean, or in the Arctic). It also motivates long-term or global observation programs to better evaluate climate and climate models over time, for example, with better satellite instruments to measure clouds.

9.3 Predicting the Future: Forecasts Versus Projections

One of the ultimate goals of building a model is to use it for prediction of the future. **Prediction** is done in two ways, and these methods—forecasts and projections—are important to understand when considering climate model results.

9.3.1 Forecasts

A **forecast** is something that we think will occur, usually assigned a probability representing our confidence in the forecast. This is common for weather forecasts (e.g., a weather prediction may call for a 50 % chance of rain at a given place and time). Some climate predictions are forecasts: a forecast of the next season or for next year based on what we know now. For events with a long lead time, like El Niño, we can make forecasts, and some of them have pretty good skill for weeks or even months in advance. For forecasting, the present state (initial condition) almost always has an impact on the forecast.

9.3.2 Projections

When we are predicting climate over long timescales, we are really talking about a **projection**. Why is a projection different from a forecast? A projection is usually dependent on things that we do not know about the future. For a weather forecast, we can assume we know all the important things that can force the weather on the scale of a few days: the composition of greenhouse gases in the atmosphere, for example. There may be important uncertainties in the fine-scale distribution, but the broad emissions are known. But in 50 years, what will the level of carbon dioxide be? That depends on what humans do, and thus we must estimate important parts of the system. We do this by constructing **scenarios** of the inputs needed for a coupled climate model: those things that force the climate system, such as the solar output or the composition of the atmosphere. Forcing usually implies effects on climate that are outside the model, like the sun. Solar output changes slightly over the course of the 11-year solar cycle, and we can estimate the change in output based on past solar cycles. Greenhouse gas emissions occur from natural sources, and these can be projected forward, but they also come from human emissions. How do we know what they will be? That requires coming up with estimates of the future evolution of the climate system forcing, and this is done with scenarios.

Scenarios are used to specify uncertain future inputs. When models are run with forcing from these scenarios, the results are not forecasts, but projections of the future, given assumptions about what might happen. The key is that the assumptions that impact the projection are outside of the realm of the model. Most climate models do not try to predict the human emissions of greenhouse gases. Integrated assessment models however (see Chaps. 7 and 8) try to predict human emissions. But these models also depend on scenarios. An integrated assessment model may generate emissions from economic activity, but even that is dependent on a scenario, of population growth, for example.

The common method is to have a series of projections spanning what we think are the possible states. For current climate models, it is common to have several scenarios of future emissions of different gases, based on assumptions about human systems, for example, economic growth and development. Each scenario used to force a climate model results in a projection. It is not really a forecast. The range of possible outcomes (projections from probable scenarios) are broadly the forecast: what we think will happen. So mostly we are considering climate projections of the future dependent on specific assumptions (scenarios) about what might happen. It is important in using models to be clear about what is imposed by scenarios and what the model is calculating. Otherwise, you might end up comparing two model projections that are different because of scenario differences, not model differences.

9.4   Applications of Climate Model Evaluation: Ozone Assessment

Perhaps the most comprehensive evaluation of climate models is conducted as part of the climate science assessment by the Intergovernmental Panel on Climate Change (IPCC). Model simulations are run by many modeling centers and evaluated by teams of scientists against observations. The IPCC scenarios are discussed more in Chap. 10, and the results are discussed in Chap. 11, with further application examples. But climate model simulations are also used to project the evolution of the stratospheric ozone layer, which has been damaged by inadvertent emissions of chemical refrigerants containing chlorine. When these chlorine containing molecules break down in the upper atmosphere (the stratosphere), the chlorine acts to destroy stratospheric ozone, increasing the penetration of ultraviolet light to the surface. The reactions occur most readily on the unique surface of clouds in the stratosphere that occur only in polar regions, mostly in the Antarctic, giving rise to a springtime (September) ozone deficit or ozone hole.

Climate models are evaluated for their ability to be able to reproduce the conditions for the distribution of ozone, and their chemistry. A comprehensive report was produced in 2010 for the analysis of the different climate models.[4] In particular, the report focused on evaluation of critical processes in the models and how they were represented compared to observations. For assessing the impact of chlorine species on ozone at high latitudes, several different processes need to be represented. First, the model must simulate the actual chemical reactions and the distribution of chlorine. But the presence of clouds in the stratosphere in the Antarctic spring is also necessary, and these cloud processes and their distribution were evaluated. The clouds are dependent on the water vapor and temperature environment. At each step, different observations were used to analyze the models. The results indicated a few models that had processes that were incorrect (wrong reactions with chlorine, for example), or that had better or worse cloud distributions. This information was used in the 2010 scientific assessment of ozone depletion, particularly in the executive summary of the assessment[5] to "select" model projections and limit the projections shown to the models shown by evaluation to have the correct representation of key processes.

9.5 Summary

Evaluation of models is targeted for a purpose: What are models good for and why? Models that are good for one purpose may not be good for other purposes. Understanding the uncertainty and, hence, the utility of a model—any model, but especially climate models—requires extensive testing against observations.

Evaluation of climate models requires some fundamental understanding of the observations themselves. What is the uncertainty and accuracy in the observations? As we discuss in detail in Chap. 10, knowing the uncertainty in our observations is an important part of being able to evaluate the uncertainty in models.

Evaluating climate models is done in many ways but is often done similar to weather models. Models are evaluated on their representation of past events, either a single event or a statistical series of events. Evaluation of climate models can also show where critical processes need to be better understood to constrain climate model projections. For example, since cloud processes and responses to environmental changes are uncertain, better representation and evaluation of clouds in climate models is critical.

---

[4]See Erying, V., Shepherd, T., & Waugh, D., eds. (2010). *SPARC Report on the Evaluation of Chemistry-Climate Models*. SPARC Report 5, WCRP-132, WMO/TD-1526. Stratospheric Processes and Their Role In Climate, World Meteorological Organization, http://www.sparc-climate.org/publications/sparc-reports/sparc-report-no5/.

[5]World Meteorological Organization. (2011). "Executive Summary: Scientific Assessment of Ozone Depletion: 2010." In *Scientific Assessment of Ozone Depletion: 2010, Global Ozone Research and Monitoring Project–Report No. 52*. Geneva: Switzerland: Author.

Predicting the future is often more of a projection than a forecast. A projection is dependent on outside factors, such as emissions dependent on economic growth and population. Thus, it is not truly a forecast because the outcome is dependent on factors outside of the model. This is a source of uncertainty that is partially outside of our knowledge of the climate system, and independent of our ability to design climate models to predict the system.

Understanding uncertainty is one of the keys to prediction, the subject of Chap. 10.

Key Points

- Proper evaluation of models requires observations and estimates of observational uncertainty.
- Observations also contain uncertain models.
- Climate models can be evaluated on many past events, or even a single past event.
- Projections of the future depend on scenarios that force climate models.

Chapter 10
Predictability

This chapter focuses on prediction of future climate and how models are used to generate predictions. Prediction includes both projections (estimates given scenarios) and forecasts (estimates given the current state of the climate), as discussed in Chap. 9. Prediction can occur over different timescales. One of the key aspects of prediction is that projections and forecasts are often not useful unless they come with an estimate of uncertainty. So we spend a good deal of space in this chapter trying to understand and characterize uncertainty.

We hear predictions about the future all the time: from sporting events to the weather. But generally a prediction is not useful unless we understand how certain it is. Suppose, for example, the forecast high temperature tomorrow is going to be 5 °F (3 °C) above normal (the average long-term mean or climatology for the date). If the uncertainty on that estimate is 10 °F (6 °C), is the forecast really that useful? If you hear of a "chance" of rain, you really need to know what chance (probability): 10 % is a lot different from 90 %. On the other hand, we can be pretty certain the sun will rise tomorrow at the predicted time (whether we can see it or not). So uncertainty becomes critical to the idea of predictability.

This is especially true for climate and climate change. Climate models can provide projections of the future. Scientists have been issuing projections for a generation or more (30+ years) of future climate using climate models (see below). A common prediction is the global average temperature response to a given forcing of the system. This forcing is often a carbon dioxide (CO_2) level of 560 parts per million, or twice the pre-industrial value of 280 parts per million. Climate model predictions of the climate response to this "doubled CO_2" forcing have not changed much in 30 years.[1] But the models have. So what have we learned? We have learned a lot about uncertainty. Incidentally, if you look at the historical temperature record for the past 30 years since many of the forecasts were published, most of the

---

[1]The basic understanding of what will happen to global temperature has not changed much (between 1 and 4 °C global temperature change for doubling CO_2 concentrations in 30 years; see Charney, J. G. (1979). *Carbon Dioxide and Climate: A Scientific Assessment.* Washington, DC: National Academies Press; and Houghton, J. T., Meira Filho, L. G., Callander, B. A., Harris, N., Kattenberg, A., & K. Maskell, eds. (1996). *Climate Change 1995: The Science of Climate Change.* Cambridge, UK: Cambridge University Press.

© The Author(s) 2016
A. Gettelman and R.B. Rood, *Demystifying Climate Models,*
Earth Systems Data and Models 2, DOI 10.1007/978-3-662-48959-8_10

forecasts have been correct or close to correct.[2] But *correct* and *incorrect* also depend on the degree of uncertainty. A forecast of 5 °F above normal, when the actual temperature is 8 °F above normal, is incorrect if the "uncertainty" in the forecast is 2 °F (a range of 3–7 °F above normal), but the same forecast is correct if the uncertainty is 4 °F (1–9 °F above normal). The small changes over the past 30 years are not enough to sufficiently constrain the future. The models are broadly "correct."

In this chapter, we review key uncertainties in model projections (and forecasts) of climate change. We start by trying to characterize and classify uncertainty, which depends on the physical system and the particular set of problems as well as the time and space scale of climate projection. We then discuss methods for using models to understand uncertainty with multiple simulations and models.

10.1   Knowledge and Key Uncertainties

What do we know about climate? Or, perhaps more important, what do we *think* we know, and what do we *not* know about climate? Uncertainty in prediction is tied to several different aspects of the system: (a) the physics of the problem, or how constrained it is; (b) the variability in the system, which is related to the underlying physics; and (c) the sensitivity of the system to changes that might occur. With regard to climate, we discuss each of these aspects in turn.

10.1.1   Physics of the System

Uncertainty depends in part on the underlying physics of the problem. Some problems are better constrained than others, and this affects our ability to predict them. In climate simulation, the global average temperature is often used as a metric. This implies an average over time and over space. The global average temperature is a fairly well-constrained number: Global averages of the incoming and outgoing energy at the top of the atmosphere and assumptions about the ocean heat uptake allow a pretty good estimate of the global average temperature. This is because the physics of energy conservation are fairly straightforward. Energy comes in from the sun. Some is reflected and some is absorbed by the surface, and then some is radiated back. There are complex constraints on how much energy is absorbed, is reflected, or remains at the surface, mostly related to cloud processes

---

[2]For an example of a forecast nearly 30 years old, see Hansen, J., Fung, I., Lacis, A., Rind, D., Lebedeff, S., Ruedy, R., et al. (1988). "Global Climate Changes as Forecast by Goddard Institute for Space Studies Three-Dimensional Model." *Journal of Geophysical Research: Atmospheres, 93* (D8): 9341–9364.

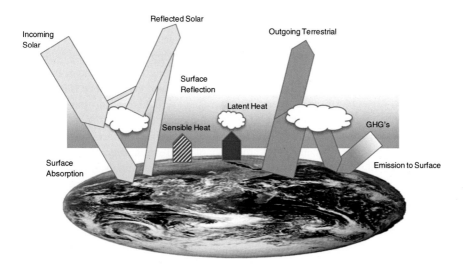

Fig. 10.1 Energy budget. Solar energy, or shortwave radiation (*yellow*) comes in from the sun. Energy is then reflected by the surface or clouds, or it is absorbed by the atmosphere or surface (mostly). The surface exchange includes sensible heat (*red striped*) and latent heat (*blue*, associated with water evaporation and condensation). Terrestrial (infrared, longwave) radiation (*purple*), emitted from the earth's surface, is absorbed by the atmosphere and clouds. Some escapes to space (outgoing terrestrial) and some is reemitted (reflected) back to the surface by clouds and greenhouse gases (*GHGs*)

and the ocean circulation, as illustrated in the energy budget diagram (Fig. 10.1).[3] But the global energy flows have limits, and the physics (conservation of energy) are fairly certain. Some of them offset each other as well. Nonetheless, we cannot predict the year-to-year variations very far in advance (more on this later).

There are other aspects of climate and efforts at prediction that are not as certain and do not have strong constraints. Prediction of regional distributions of rainfall, for example, is not as well constrained by total energy budgets. In addition to local evaporation and precipitation, air motions bring moisture sources into a region from various and changing direction. The total available energy in a region for precipitation depends on the global atmospheric circulation. It may also depend on details of cloud processes. So predicting the distribution of rainfall at any location is much less certain than global quantities due to the basic physical constraints.

---

[3]For an overview of the earth's energy budget, see Chap. 2 of Trenberth, K. E., Fasullo, J. T., & Kiehl, J. (2009). "Earth's Global Energy Budget." *Bulletin of the American Meteorological Society, 90*(3): 311–323.

10.1.2   Variability

Another aspect of uncertainty related to the underlying physics is the level of variability. Recall Fig. 9.2, with two distributions having the same mean but different variance (represented by the **standard deviation**). If the variance is high, we may not be very good at prediction, since the quantity is highly uncertain. Global average temperature is an example of a distribution with low variance, whereas regional rainfall is a distribution with high variance. Higher variances are harder to limit and predict, often because our knowledge of the distributions is low. It is hard to tell if a **signal** (change in mean, for example, or change in variance) is significant if there is a large amount of variability. The local temperature example in the introduction to this chapter is an example of this. A temperature deviation from a mean by 4 °F (2 °C) is hard to detect if the variance is large, usually measured as a standard deviation from the mean (see Chap. 9). The probability of the deviation's being "significant" depends on how variable the distribution is, so larger variability (standard deviation) yields less certainty.

The concept of variability is used to help understand the difference between a "signal", and "noise."[4] The signal is a change in the mean: a trend in the climate over time. The noise is the variability that occurs around the mean, such as the variability of a given climate metric from year to year. The larger the variability (noise), the harder it is to see the signal. Higher variability typically occurs on smaller space and time scales. The annual average temperature of a particular location is much more constant than the daily temperature, and the global average temperature varies far less (daily or annually) than does the average temperature in any particular location or region.

10.1.3   Sensitivity to Changes

Finally, and related to the concept of distributions, impacts often depend on the sensitivity to changes of a distribution. Sensitivity has two aspects. First, we may be worried about extreme events, which are low-probability events at the tail of the distribution. Changes to extreme events are hard to predict because they are low-frequency events, and they have high variability. This is another aspect to uncertainty in prediction that stems from the nature of the physics of climate.

Second, we may worry about events that have thresholds. For example, if the temperature is 68 °F (20 °C), and there is a 1–2°(F or C, F is bigger) uncertainty, that might not matter much. But if the temperature is 32 °F (0 °C), being wrong by a few

---

[4]For a good popular treatment of the signal and noise in statistics and in climate modeling, see Silver, N. (2014). *The Signal and the Noise: Why So Many Predictions Fail—But Some Don't.* New York: Penguin. Chapter 12 is all about climate science and climate prediction.

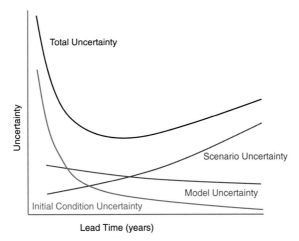

Fig. 10.2 Types of uncertainty. Uncertainty is illustrated here as a function of lead time. Initial condition uncertainty is shown in *green*, model or structural uncertainty in *blue*, scenario uncertainty in *red*, and total uncertainty (the sum of three uncertainties) in *black*. Initial condition uncertainty is large initially, then shrinks, and scenario uncertainty grows over time. *Source* Based on Hawkins and Sutton (2009)

degrees in either direction makes a big difference: If ice and snow are present and remain, then the weather situation is a whole lot different than if a cold rain falls.

10.2   Types of Uncertainty and Timescales

In addition to the physics of a problem contributing uncertainty (which we characterize later in the discussion of model uncertainty), there are uncertainties related to the timescales of prediction that have profound implications for climate modeling. Prediction has different sources of uncertainty on different timescales. These can be divided into uncertainty in predicting the near term (initial condition uncertainty), predicting the next few decades (model uncertainty), and predicting the far future (scenario uncertainty).[5] The categories are the same as introduced in Chap. 1. Understanding uncertainty and what it means is a critical tool for evaluating climate models, that is, for understanding whether they are fit for the purposes of prediction for which the models are used. The different types of uncertainty are illustrated in Fig. 10.2.

---

[5]The definitions follow Hawkins, E., & Sutton, R. (2009). "The Potential to Narrow Uncertainty in Regional Climate Predictions." *Bulletin of the American Meteorological Society, 90*(8): 1095–1107.

10.2.1   Predicting the Near Term: Initial Condition Uncertainty

Short-term climate prediction commonly means predicting the climate from the next season to the next decade or two. In many respects, it is similar to weather forecasting. In the climate context, a decade is "short term." In some sense, the short term is characterized by the period over which the initial state of the system matters: The weather tomorrow depends on the weather today. Other examples include routine seasonal forecasts of expected high and low rainfall one season in advance. Broadly, we are defining "short term" as the period over which the current state of the system matters, and when we can predict the state of some of the internal variability of the climate system.

As described earlier, prediction depends on the problem. Some parts of the system are more predictable than others. In some regions, rainfall in the next season depends on things that are predictable several months in advance, like ocean temperatures. For example, current and predicted temperatures of the tropical Pacific Ocean provide predictive skill for rainfall in western North America.[6] In other places, like Northern Europe, extremes of precipitation or temperature (both high and low) are usually functions of **blocking events**, which are weather patterns that persist but are predictable only a week or so in advance. The extent (persistence) of blocking patterns is difficult to predict at all.

Estimates of the near term (whether a week or a season in advance) made with skill can be valuable in preparation. For example, El Niño commonly brings wetter conditions to Southern California, and when a strong El Niño is developing, precautions are often taken to deal with mudslides, especially in regions recently affected by fire. Forecasts of impending tropical cyclone impacts 24–72 hours in advance enable evacuations and staging of rescue and disaster relief.

But high variability that makes predictions uncertain means that they are less useful. As a counterexample, El Niño also alters the frequency of occurrence of tropical cyclones in the Atlantic (hurricanes), because the upper-level winds in the Atlantic blow the opposite way from usual (and in the opposite direction to surface winds). However, while storms are less frequent, even a single storm in an El Niño year can be damaging. In 1992 after a large El Niño, there were only four hurricanes that year in the Atlantic and one major storm (average is six hurricanes and three major ones), but that single storm was Hurricane Andrew, the second most devastating (in financial terms) storm in recent U.S. history. Because the variability of tropical cyclones is extremely large, and the small-probability events are devastating, the improvement in understanding and forecasting may not be that valuable for prediction.

Near-term prediction is very much focused on the current state of the system: What El Niño will do in 6 months depends strongly on the state of the system now.

---

[6]The tropical Pacific temperatures are due to El Niño; see Chap. 8.

It is very similar to trying to predict the weather a few hours or day ahead. Most of the uncertainty is in the initial state of the system, called **initial condition uncertainty**. As an example, the El Niño forecast depends strongly on the current state of the atmosphere and ocean; it does not depend on the CO_2 concentration in the atmosphere in 6 months.

Initial condition uncertainty is a common problem between weather and climate prediction: The climate system is usually sampled sparsely with uncertain observations, and we do not know key things to help us "constrain" the system to forecast the weather more than a few days in advance. If we had more and better information (observations of temperature and wind) in the right places, we would be able to reduce initial condition uncertainty.

For climate prediction (e.g., What will the distribution of weather states be next year, in the year 2020, in 2080, or 2200?), initial condition uncertainty dominates over other uncertainty out to decadal scales. This is because there are long-term patterns in the climate system, mostly from the transfer of heat to and from the deep ocean. At scales beyond a few months, it is the evolution of the ocean that governs how the system responds. Figure 10.2 illustrates that initial condition uncertainty is very large and dominates the uncertainty initially but averages out over long periods of time. The weather tomorrow is dependent on today, and El Niño next year is dependent on this year. But the weather and El Niño's state in 2080 is not dependent on today.

10.2.2   *Predicting the Next 30–50 Years: Scenario Uncertainty*

As the example of the weather or El Niño suggests, there are certain scales for which we know the state of how the system is "forced": Tomorrow's weather does not depend much on the greenhouse gas concentration. Over the course of a few years, the earth's energy budget does depend on the greenhouse gas concentration. But this concentration is fairly well known. As we described in Part II, there are sources and sinks of carbon, and these balance to leave a reservoir of carbon in the atmosphere (see Fig. 7.6). The CO_2 concentration (and that of methane) changes slowly, with variable growth rates, but in the near term—next year, or in the next 10 years—we have a pretty good idea of what will happen. The total use of fossil fuels also changes slowly over time. How much energy did you use in your house this year? How much gas did you use either in your own car, or as part of sharing a ride or using public transportation? Absent major life changes (new house, new job, new car), you can probably project that your use next year will be similar. So it goes for societies in general. Figure 10.3 illustrates the global emissions of carbon. It is a line that trends up a little bit from year to year. You can make a good estimate of next year's emissions from this year's, and projecting forward with a line is not too hard.

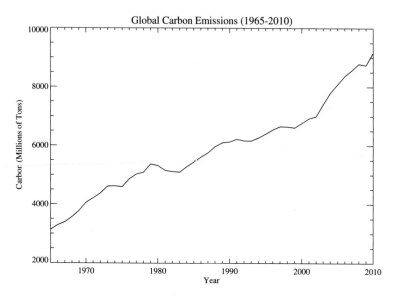

Fig. 10.3 Recent CO_2 emissions. Total global carbon emissions over time from 1965 to 2011 in millions of tons. Data from the NOAA Carbon Dioxide Information Analysis Center (NOAA)

But over longer periods of time, things change. What will your personal energy use be in 10 years? That depends on where you live, your job, your house, if you have a family. It also depends on how you will use energy. For example, next year you might put some more efficient lights in your house. But in 10 years you might have an electric car, or you might telecommute, or you might have solar panels, and it might be the same house or a different house. So the uncertainties on your energy use increase. And the uncertainties on your carbon emissions get larger. For instance, you will probably still have a refrigerator. We can probably guess it will use a similar amount of energy as today (maybe you will have the same refrigerator), but maybe you (or your utility) will get the energy for that appliance from solar or wind or gas rather than coal. Those changes mean more variability in the future. Figure 10.4 shows the same data as Fig. 10.3, but now from 1850 to 2011. Now the use is harder to project into the future. If the plot stopped in 1900 or 1950, the "projection" might be very different from what you would assume today. The mix of fuels would be different: a projection of petroleum use in 1970 would be very different from one made today. And you can guess in the future that it might also be very different, for any number of reasons.

Figure 10.5 shows an example of a plot that does *not* continue to increase. Figure 10.5 is of another greenhouse gas emission: CCl_2F_2, dichlorodifluoromethane, also called CFC-12,[7] a refrigerant used in home and industrial coolers from refrigerators in homes to air-conditioning units in cars or buildings. It also

---

[7]CFC-12 data from McCulloch, A., Midgley, P. M., & Ashford, P. (2003). "Releases of Refrigerant Gases (CFC-12, HCFC-22 and HFC-134a) to the Atmosphere." *Atmospheric Environment, 37*(7): 889–902.

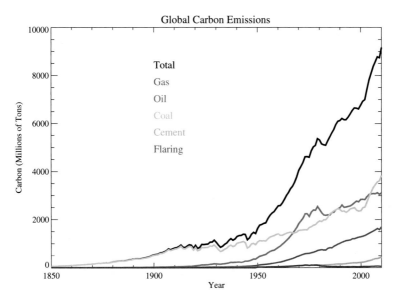

Fig. 10.4 CO_2 emissions by sector. Global carbon emissions by sector from 1850 to 2011 in millions of tons. Total emissions are the sum of the other sectors. Gas (methane, CH_4), Oil and Coal sectors are burning of these fossil fuels. Flaring is burning methane at oil wells, and CO_2 from cement production occurs due to the chemical process and the fossil fuels used for energy to drive the process. Data from the NOAA Carbon Dioxide Information Analysis Center (NOAA)

happens to contain chlorine (Cl), and when it decomposes in the upper atmosphere this chlorine destroys stratospheric ozone, resulting in the Antarctic ozone hole. As a result of this effect, CFC-12 use has effectively has been banned. Again, projecting the emissions of this substance in 1960 or 1980 into the future would be very different from what actually happened. In the case of CFC-12, it was regulated under the 1986 Montreal Protocol, and eventually production had stopped over most of the planet.

You can see where this is going. Over short periods of time, the basic inputs to a climate model—concentrations of greenhouse gases—are fairly predictable. But over longer periods of time, due to external factors (political oil shocks, new technologies) or regulation (limits or bans on productions or emissions), the inputs can change dramatically. For the historical period, whether for carbon or CFCs, we have a pretty good handle on emissions. However, what does this mean for the future? Will oil use rise, as in Fig. 10.4? Will the Chinese have as many cars per person as in the United States? Or is oil use going to look more like the curve in Fig. 10.5, because either everyone is going to have solar panels on their roofs or we run out of economical supplies of oil.

These are not questions for climate models to answer. They are beginning to be questions for integrated assessment models (see Chap. 8). But the answers (how much oil will be used, and how much CO_2 be will emitted by humans in the year 2080) are highly uncertain. They depend on myriad socioeconomic choices. They are not so much predictions as projections and they represent scenarios. Oil use in the future is dependent on a lot of interrelated things: population, population density,

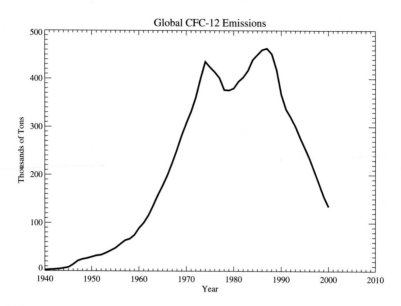

Fig. 10.5 Global CFC-12 emissions from 1930 to 2000 in thousands of tons. Emission controls began in 1986. Data from McCulloch et al. (2003)

available economical supplies of oil, new technologies, and how all these factors interact. One of the problems of scenarios is predicting the application and spread of **disruptive technologies**. Integrated assessment models are better at predicting the improving efficiency of refrigerators, or power plants, or solar panels. They are less able to predict the impact of smart phones or creative financing for solar panels.

It should be clear that **scenario uncertainty** (the red line in Fig. 10.2) grows over time and the inputs to a climate model (emissions of gases that affect climate) become highly uncertain. This occurs for timescales longer than a human genera-tion (20–30 years), which also corresponds to the depreciation lifetime of many fixed investments (like a power plant). The major drivers of these uncertainties are slow-accumulating things, like global population, or commonly used energy sys-tems. They are usually subject to the results of different societal decisions: popu-lation growth in China, gas extraction technologies in the United States, economic growth in developing countries (Brazil, India, China), and disruptive technologies (internet, global manufacturing, and so on).

One way to look at this is to look at past projections of the present. Figure 10.6 is a reprint from a 2014 energy policy report from the Office of the President of the United States. The report is political, but the data presented in the figure are simply the consumption of gasoline for vehicles in the United States. In 2006, it was projected that in 2013 the United States would use 10 million barrels (550 million gallons, or about 2 billion liters) of gasoline per day. The projection for 2030 was 12 million barrels per day. But the actual value for 2013 was closer to 8 million barrels per day, and the 2030 forecast estimated in 2014 is only 6 million barrels per day. These are big numbers, with big implications. If we wanted to be 40 % below

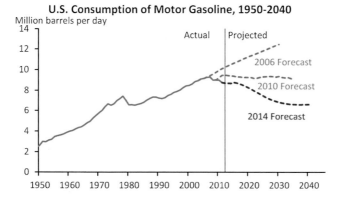

Fig. 10.6 Changing forecasts. Historical and projected U.S. consumption of gasoline in millions of barrels per day. Different forecasts started from different years are shown with the *dashed lines*. *Source* Executive Office of the President, 2014, "The All of the Above Energy Strategy as a Path to Sustainable Economic Growth," Fig. 2.2b. Data from Energy Information Agency

2006 consumption in 2030, that would take extreme measures for the 2006 forecast, but that would be the expected scenario in 2014 (without doing anything). So scenarios are forecasts as well, and they are often wrong.

Projections of economic and social indicators span a range of predictability. It is probably pretty easy to estimate population now and for the next few years. But economic growth or prices of commodities? Or use of gasoline? Economic growth estimates are rarely even known until well after the fact (the U.S. government revises all growth figures a few months after they are issued). These errors all compound to become part of the uncertainty in the input scenarios used to drive climate models.

So how do we deal with these scenario uncertainties in climate projections? These are the uncertainties in the forcing of climate. In the absence of making a prediction about what society might do, generally social scientists and economists estimate what is possible, and a range of possibilities (scenarios) are derived. The inputs to climate models are typically outputs of integrated assessment models, driven more by a story translated into quantitative inputs. These scenarios have evolved in complexity over time, but the values and spread have not changed that much. But they are not forecasts, because they depend on human policy decisions and choices (like the CFC-12 curve in Fig. 10.5).

Most climate models now use commonly developed scenarios.[8] Figure 10.7 illustrates the latest set of scenarios (**Representative Concentration Pathways,** or RCPs) used to force most climate models.[9] Figure 10.7 illustrates the scenarios in

---

[8]For a background on scenario development, consult Nakicenovic, N., & Swart, R., eds. (2000). *Special Report on Emissions Scenarios*. Cambridge, UK: Cambridge University Press. See also https://www.ipcc.ch/pdf/special-reports/spm/sres-en.pdf.

[9]A full description of the RCP scenarios is at Van Vuuren, D. P., et al. (2011). "The Representative Concentration Pathways: An Overview." *Climatic Change, 109*: 5–31.

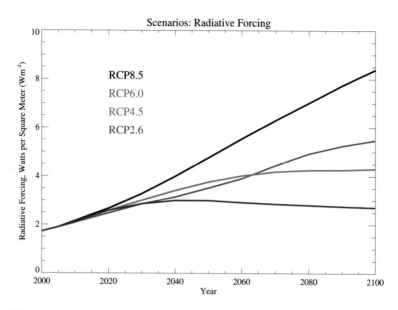

Fig. 10.7 Radiative forcing (in Watts per square meter) over time from 4 representative concentration pathways (*RCPs*) illustrating the most common scenarios used to force climate models. RCP8.5 (*black*), RCP6.0 (*blue*), RCP4.5 (*green*), RCP2.6 (*red*)

terms of the radiative forcing of climate in watts per square meter (Wm^{-2}) of the earth's surface. Recall that a watt is the same as the watt in lightbulbs: 60 Wm^{-2} is the energy of a typical incandescent bulb, and the solar input is about 160 Wm^{-2}.[10]

Each RCP scenario was designed to reach a specific radiative forcing target from greenhouse gases by a particular date, based on different predictions of what society might do. The levels of CO_2 and other greenhouse gases in the atmosphere that create the forcing are derived for each RCP (Fig. 10.8). CO_2 concentrations are in parts per million (ppm). One part per million means one molecule of CO_2 for every million molecules of air. The current concentration of CO_2 is about 400 ppm, and the level in 1850 before most industry developed was 280 ppm CO_2. That is a 40 % increase. Note that these scenarios are possible futures, just like the different forecasts in Fig. 10.6. We have been on the black curve (high emissions), but Fig. 10.6, and others like it, suggest that we might be transitioning to one of the more moderate curves.

Using common scenarios implicitly says that the scenarios are a major driver of uncertainty, and we need to compare models with common scenarios. Otherwise, the uncertainty due to the scenarios would be hard to assess.

As we shall see, a central theme of future climate prediction is to realize that most of the uncertainty lies not in the models themselves, but in the scenario for emissions (i.e., in the scenario uncertainty). The climate of the future is dominated

---

[10]Trenberth et al., "Earth's Global Energy Budget.".

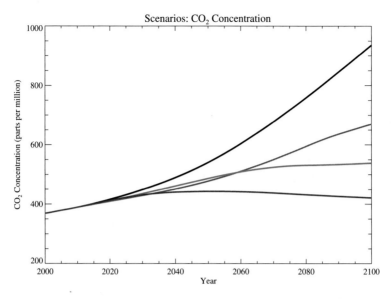

Fig. 10.8 CO_2 concentration scenarios. Projections of atmospheric CO_2 concentration (parts per million) over the 21st century from different representative concentration pathways (*RCPs*) used for climate model scenarios. RCP8.5 (*black*), RCP6.0 (*blue*), RCP4.5 (*green*), RCP2.6 (*red*)

by scenario uncertainty (the red line in Fig. 10.2), and that uncertainty is from human systems. In the end, climate is our choice, even if it is simply to make no choice about our emissions and continue present policies.

10.2.3   Predicting the Long Term: Model Uncertainty Versus Scenario Uncertainty

Thus far we have discussed sources of error in climate projections for the far future (beyond several decades) and in the near term (within a decade or two). Throughout this entire period there is a constant source of uncertainty, and this is what we typically think of as the uncertainty in a climate model: the model uncertainty, or the structural uncertainty in how we represent climate (the blue line in Fig. 10.2).

Model uncertainty is our inability to represent perfectly the coupled climate system. It is present in representing each of the components, and each of the processes within the system. The uncertainty has many different forms, and different values. Some uncertainties are large and some are small, mostly related to the physics of the processes. We have discussed a lot of these physical and process uncertainties in detail. Many uncertainties stem from the scales that models must resolve: Small-scale processes that are highly variable below the grid scale of a model are difficult to resolve or approximate. In the atmosphere, clear skies are easier to understand and represent than cloudy skies. In the land surface, variations in soil moisture and vegetation at the small scale interact with moisture and energy fluxes to make things complex.

Complexity arises due to nonlinear effects. In a nonlinear process, the average of results is not the same as if the inputs are averaged. As a simple example, let's say that the rate of evaporation (R) is the square of the soil moisture ($R = S^2$), where the moisture (S) varies from $0 \rightarrow 10$. If there are two equal-sized areas of a grid box, one with soil moisture of $S = 0$ and one of $S = 10$, then the average soil moisture $S = 5$. But calculating the evaporation from the average yields $R = 5^2 = 25$, whereas calculating the rate from each part and averaging (0 and $10^2 = 100$, divided by 2) yields 50: double the rate.

Thus there are many sources of model uncertainty. In the present context of trying to understand the sources of uncertainty, note that model uncertainty is constant over time: It does not increase. Because in the future there are better large-scale constraints than small-scale constraints, uncertainty in the future might even decrease.

Model uncertainty can be broken into parametric and structural uncertainty.[11] **Parametric uncertainties** are those that arise because many processes important for weather and climate modeling cannot be completely represented at the grid scales of the models. Therefore, these processes are parameterized, and there are uncertainties in these representations. An iconic parameterized process is that associated with convective clouds that are responsible for thunderstorms. Thunderstorms occur on much smaller spatial scales than can be resolved by global climate models. Therefore, the net effect of thunderstorms on the smallest spatial scales that the climate model can represent is related to a set of parameters based on physical principles and statistical relationships. Recall that the spatial scale that is resolved by a climate model is several times larger than the spatial size of a single grid point. The conflation of errors of different spatial scales, dependent on parameterizations, is difficult to quantify and disentangle.

Structural uncertainties are those based on decisions of the model builder about how to couple model parameterizations together and how they interact. One class of these errors is the always-present numerical errors of diffusion and dispersion in numerical transport schemes; that is, the errors of discrete mathematics. But a more important structural uncertainty perhaps is how the different parameterizations are put together. For example, let's say an atmosphere model has convection, cloud microphysics, surface flux, and radiation parameterizations. Typically there is a choice in how to couple parameterizations. Either all the parameterizations are calculated in parallel from the same initial state, called **process splitting**. Or each one could be calculated one after the other in sequence, called **time splitting**. Typically in a model with time splitting, the most "important" processes are estimated last. For weather models, this is usually cloud microphysics to get precipitation correct. For climate models, this is usually radiation. In either case, the last process is the most consistent with the sequential state. Either method, however, may create a structural uncertainty: If clouds and radiation are calculated

---

[11]Tebaldi, C., & Knutti, R. (2007). "The Use of the Multi-Model Ensemble in Probabilistic Climate Projections." *Philosophical Transactions of the Royal Society A: Mathematical, Physical and Engineering Sciences, 365*(1857): 2053–2075.

separately in a process split model, then the radiation may not reflect the clouds, and the temperature change from heating or cooling may not add up correctly. These are structural uncertainties from coupling processes. The same uncertainties come from coupling the physical processes to the dynamics in the atmosphere or ocean, and from coupling different components.

The result is an uncertainty graph that looks like Fig. 10.2. Internal variability, which is defined from the initial conditions (green line), is the largest contribution to short-term variability (the next tropical cyclone or blocking event or ENSO does not care much about the level of greenhouse gases), with a significant component of model variability (blue line). Over time (a few cycles of the different modes of variability), this fades, and the uncertainty is dominated by structural uncertainty in the model (blue line). But over long time periods, where the input forcings are themselves uncertain, the scenario variability (red) dominates. The latter break point is approximately where the level of uncertainty in the forcing (i.e., uncertainties in emissions translated into concentrations of greenhouse gases, translated into radiative forcing in Wm^{-2}) is larger than the uncertainty in the processes (expressed in radiative terms, Wm^{-2}). For climate models, the largest uncertainty is in cloud feedbacks. The uncertainty in cloud feedbacks is about 0.5 Wm^{-2} per degree of surface temperature change. So far the planet has warmed about 1–2 °C so the uncertainty due to cloud forcing is about 1 Wm^{-2}. This uncertainty would indicate that scenario uncertainty will dominate model uncertainty when the uncertainty in future radiative forcing between scenarios reached about 1 Wm^{-2}. In Fig. 10.7, this would be about 2040–2050 and beyond. The practical significance of this situation will be demonstrated when climate model results are discussed in Chap. 11. But first a few words about using models together in ensembles.

10.3 Ensembles: Multiple Models and Simulations

Different strategies can be used to better quantify the different types of uncertainty in climate model projections. One of the best, which can be used for quantifying all three types of uncertainty, is to run **ensembles**, or multiple simulations. These simulations are typically done to gauge one kind of uncertainty, but they can be appropriately mixed and matched to understand all three types of uncertainty.

The use of ensembles is now also common in weather forecasting. Weather forecasting suffers from model uncertainty and especially initial condition uncertainty, but usually not scenario uncertainty as we have described it. However, sometimes external events such as fire smoke can significantly affect the weather. In weather forecasting, ensembles of a forecast model are typically run in parallel: multiple model runs at the same time. Then the spread of results of these parallel runs is analyzed. The simulations usually differ from each other by perturbing the initial conditions (slight variations in temperature, for example) and/or by altering the parameters in the model. Sometimes forecasters consult different models, creating another sort of ensemble. Altering initial conditions tests the initial condition uncertainty, whereas altering the model or the model parameters within a model tests model uncertainty. Ensembles are one way to assign probabilities for forecasts: An 80 % chance of rain, for example, comes directly from an ensemble with 10 members, equally likely, where 8 of the 10 predict rain. Ensemble forecasting provides an estimate of uncertainty: If you run a forecast model once and get an answer (sunny or rainy), how do you know how good that answer is (what is the uncertainty)? If you change the inputs or the model slightly and you get the same answer every time, it is probably pretty robust. If the answer changes in half the ensemble "members," then maybe it is not so robust.

The same methods can be applied to climate forecasting,[12] to estimate model uncertainty and initial condition uncertainty. In addition, different scenarios can be used to test the scenario uncertainty.

To start, one way of testing initial condition uncertainty in a climate model projection is to perform several different simulations of the same model, with the same forcing (scenario) but different initial conditions. This is often done to assess the internal variability of the model. For example, start up a model in 1850, use observations of how the atmosphere, sun, and earth's surface have changed, and see what the results are for the present day. Because of lots of different modes of variability in the system, some lasting decades, it is important to get a complete sample of the possible states of these modes.

Model uncertainty can then be added either by perturbing the model or by using multiple models. Technically, each perturbation to the model parameters is a different model, albeit one that is quite close to the original. But using different models developed quasi-independently in different places by different people provides a pretty good spread of answers. This can show the range of different possible states given a set of input parameters and a particular scenario. One must be careful because many models are not independent: They share common components and common biases.

Finally, models (either the same or different, probably both) can be run with different forcing scenarios, and the results compared. This is commonly done for the future, but it is also effective to do for the past. A nice illustration of this technique (Fig. 10.9) comes from the 2007 Intergovernmental Panel on Climate Change (IPCC) report[13] (see Chap. 11), where a set of models was run for two historical scenarios, one including human emissions of greenhouse gases and particles, and one using only natural sources. The results of the two different ensembles of models are different, and only the one with human emissions matches temperature observations over the 20th century.

In Chap. 11, we show combinations of these ensembles. For example, several different simulations of a single model or multiple models with the same scenario

---

[12]Knutti, R., Furrer, R., Tebaldi, C., Cermak, J., & Meehl, G. A. (2010). "Challenges in Combining Projections From Multiple Climate Models." *Journal of Climate, 23*(10): 2739–2758.

[13]Solomon, S., ed. (2007). *Climate Change 2007: The Physical Science Basis: Working Group I Contribution to the Fourth Assessment Report of the IPCC*, Vol. 4. Cambridge, UK: Cambridge University Press.

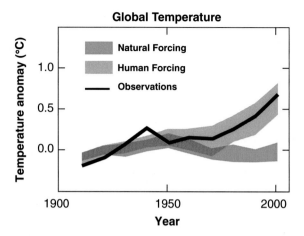

Fig. 10.9 Global average temperature anomalies from 1900 to 2000 based on climate model simulations. Only the models with natural and human emissions (*pink*) match the observations (*black*). Models with only natural emissions (*blue*) do not match observations. Adapted from Solomon et al. (2007), Summary for Policy Makers, Fig. 4. *Note* Solomon, S., D. Qin, M. Manning, Z. Chen, M. Marquis, K. B. Averyt, M. Tignor, and H. L. Miller, eds. (2007). *Climate Change 2007—The Physical Science Basis: Working Group I Contribution to the Fourth Assessment Report of the IPCC (Climate Change 2007)*. Cambridge, UK: Cambridge University Press

can illustrate the different possible climate outcomes (i.e., the different simulated changes in global average surface temperature) with a single scenario. Ensembles can be built the same way with other scenarios and compared, so that the different contributions to initial condition uncertainty (variations within a single scenario and model but with different initial conditions), model uncertainty (variations within a single scenario using different models), and scenario uncertainty (either of the first two single-scenario ensembles with single or multiple models, but now for multiple scenarios) can all be assessed, and to some extent quantified.

There is much discussion about how to evaluate and weight models in a group: If a model resembles observations of the present, is it better and should it be given more weight[14]? If a model does not meet certain tests or standards, or has known problems, should it not be analyzed? In general, one problem with weighting models is that present performance on one metric in the past is not necessarily a good indicator of skill for the future. It is the same as the statement at the bottom of reports on the history of a financial investment: "past performance may not be an indicator of future returns." Models with fundamental flaws are sometimes removed from analyses or given less weight. However, many times these "flaws" are designed to constrain the models in some way, and are often justifiable. There is still much debate in the scientific community over how exactly to construct weighted averages across ensembles with multiple models.

[14]Knutti, Reto. (2010). "The End of Model Democracy?." *Climatic Change, 102*(34): 395–404.

But broadly, there is the interesting property that the wisdom of crowds applies. If the models have random mistakes and their biases are uncorrelated, then the average of the models (the multi-model mean) tends to be better than most of the models. This is a big "if," and many models have dependencies, but in practice (and in synthetic statistical tests), because the climate system is complicated, and most models meet basic measures of representing the system (conservation of energy and mass), the multi-model mean is usually a pretty good statistic. This makes ensembles quite valuable.

One thing to remember is that there is lots of chaotic internal variability in the climate system, and the observed record is only one possible realization of that internal variability. There may or may not be an El Niño event this year, but the possibility is that there could have been. So with the real climate system, we have only one ensemble. We expect models to be different from observations in any given year, or even decade, if they are fully coupled and internally consistent. This also confounds prediction and evaluation. It is one reason why the present and the past do not fully constrain the future (see box).

Why the Present and Past DO NOT Constrain the Future

It is often assumed that a model must be able to represent the present (or past climate) correctly to represent the future. This is true. However, while necessary, it is not a sufficient condition to constrain the future. Suppose that a model represents the present for the wrong reasons. Perhaps there is a large compensation of errors: Maybe an error in the radiation code letting in too much energy is compensated for by an error in the sea-ice model that reflects too much energy. If the planet warms up, the sea ice will go away, but the error in the radiation code will remain. It is also assumed that the more accurate a model is at representing the present (or the past), the better it will be at representing the future. This is also true, but it requires that "better" representations be for those areas that matter for future climate, and this is not necessarily the case.

What does that mean in practice? Let's say you have two measures of model performance. One is based on clouds, and one is based on temperature. If model A scores 80 % on both temperature and clouds relative to observations, and model B scores 100 % on clouds but 20 % on temperature, then a simple average score says that model A scores 80 % and model B scores 60 %. But if present clouds are more important for future climate than temperature, model B may be better. In the previous example, the present-day model may score well on sea ice and lower on the radiation code, but the radiation code is more important for the future. Also, knowing the mean state today does not imply that you can predict the change in that state accurately in the future. It is usually a good indicator, but not guaranteed.

That covers why the present does not necessarily constrain the future. Because with present climate we do not know what the response to a forcing (i.e., feedbacks) will be. But what about using the past: If we know observations of the past state, can't we use them, and run models with them, to help us understand the future? Yes, we can, and we do. A good test of a climate model is to run it for situations representing paleoclimates like the last glacial maximum (peak of the last ice age), or the change from the last glacial period to today. This does give some hint as to the potential changes in the system, and whether a model can represent them.

But it does not tell the whole story, and there is currently quite a bit of confusion in the community over the value of the past. Many times, observations and models are examined to look at changes to climate and to estimate the response. If the forcing is known, then this also tells us something about the feedbacks. That is, if we know CO_2 changed a certain amount between the last ice age and today (180–280 ppm), and we know what the temperature change is $-15\ °F$ ($-8\ °C$) in the ice core record; see Fig. 3.2), can't we just back out a relationship for temperature change for the next 100 ppm of CO_2? Or more appropriately, can we calculate the radiative effect of that extra CO_2 and we then get the sensitivity of the climate system (°C per Wm^{-2})?

The problem is twofold. First, the causes and effects are different. CO_2 changes were likely a response to changes in the earth's orbit that ended the ice age. The forcing started with changes to the total absorbed radiation from the sun, and CO_2 responded to changes in the ocean circulation. Second, the sensitivity (and the feedbacks) are dependent on the state of the climate system. There was a lot more ice and snow back then and less water vapor in a colder atmosphere, so the ice albedo feedback and the water vapor feedback were likely much different than today. The water vapor feedback was likely smaller, but the albedo feedback larger. It is also worth noting that the processes and pathways are different: Cloud feedbacks, for example, may respond differently to a change in solar radiation (which primarily heats the surface) than to a change in infrared (longwave or terrestrial) radiation, which heats the atmosphere as well as the surface.

10.4 Applications: Developing and Using Scenarios

The development of scenarios is a good example of the interactions between uncertainty and the importance of understanding how models are constructed for appropriately using them. For application purposes, Representative Concentration Pathways (RCPs) were developed by using integrated assessment models. RCPs were produced by making assumptions about the desired radiative forcing at the end of the 21st century, and different integrated assessment models were used with different input assumptions. These assumptions include population and economic growth, as well as technology improvement. The implication is that each RCP with different emissions implies a different future forecast for society. RCPs with high forcing, like RCP8.5, have large emissions of greenhouse gases, but also assume a large and wealthy global population that would produce these emissions. This means that if you are using an RCP, it should be consistent with the assumptions for the application about the future of the anthroposphere. Many critics argue that the RCP8.5 is impossible to reach: The emissions are too high given what we expect about population growth.

With RCPs, the information about the societal pathway consistent with emissions is not very clear. In recognition, new scenarios are being produced using a different methodology, called **Shared Socioeconomic Pathways** (SSPs).[15] SSPs were developed to have a consistent story. They start with assumptions about the anthroposphere: population, economic growth, technology, and efficiency improvement. These assumptions are more widely available and published, and they are used in several different integrated assessment models to generate the emissions consistent with those pathways.

What this means for applications is that the appropriate SSP (or RCP) output from a set of climate models to use depends on your assumptions about the future state of humanity, and that for specific applications, not all the scenarios should be treated equally, and the application needs to be consistent with the scenario. For example, if the aim is to look at impacts on forests, then an RCP with high emissions, which might assume that many forests have been cut down, is not appropriate. The climate that results from the RCP (with fewer forests specified) may not be appropriate for looking at climate impacts on forestry. Since climate changes in currently forested regions may assume that they become another ecosystem type (like cropland), and assuming in the future that these locations still represent forests is wrong. The same caution exists for the SSPs, but these newer scenarios try to make explicit the assumptions about the assumed evolution of the anthroposphere.

10.5 Summary

Projections in climate models need to have uncertainty attached to them. Uncertainty can be usefully broken up into three categories: initial condition uncertainty (also called internal variability), model uncertainty (also called structural uncertainty), and scenario uncertainty. On short timescales, from days to decades, the initial conditions or internal state of the system matter. Aspects of the climate system are predictable on different timescales, out to a decade (when the deep ocean circulation is involved). Model uncertainty affects all timescales and is not that dependent on timescale. Model uncertainty can be divided into uncertainties in parameters

---

[15]O'Neill, B. C., et al. (2014). "A New Scenario Framework for Climate Change Research: The Concept of Shared Socioeconomic Pathways." *Climatic Change, 122*(3): 387–400.

(parametric uncertainty) and larger questions of how models are constructed and what fields they contain (structural uncertainty). On long timescales, longer than the internal modes of variability in the climate system (one to two decades), scenario uncertainty dominates climate projections. Scenarios are an exercise in predicting and making our own human future. Scenarios are often produced with models that have their own uncertainty, and scenarios have changed over time as human society changes. The history of the climate system is one way to look at a scenario, but this may not be a complete representation of the climate system.

One way of teasing out the different uncertainties using climate models is to run multiple simulations called ensembles using combinations of different initial conditions, different models, and different scenarios. We explore these results from the latest generation of models in Chap. 11.

Key Points

- Models have different sources of uncertainty, which are important at different timescales and for different problems.
- Scenarios are highly uncertain, and the future of human systems can change rapidly.
- Multiple models and multiple simulations, called ensembles, can be used to sample uncertainty.

Chapter 11
Results of Current Models

This chapter provides some perspective on current results and modeling efforts, taking into account what we have already discussed about the climate system, climate models, and uncertainty. Rather than present detailed results from climate models, which will change as model versions change, we use selected results of recent climate model simulations to characterize and frame the uncertainties we have already discussed. The goal is to understand the uncertainty in climate model predictions of the future. A prediction without uncertainty, or with the wrong uncertainty, may be worse than no prediction at all. For example, if the prediction is for a temperature of 34 °F (1 °C) with precipitation and the actual temperature is 28 °F (−2 °C), the prediction is still correct if the uncertainty is ±6 °F (3 °C) or more. But if you do not have the uncertainty range, and you assume the temperature is going to be above freezing, then you might not have planned for snow rather than rain, or for freezing rain and ice. If you knew the uncertainty was large, you would plan for snow and ice.

The discussion also provides examples of the predictions of current climate models and the level of uncertainty. First, we briefly review some of the history and organization of modeling efforts. Second, we discuss what we want to know (predict) and how to use uncertainty. Third, we review the confidence in current predictions. Our goal is to frame the uncertainty with specific examples to assist the reader in assessing models for their needs and applications.

11.1 Organization of Climate Model Results

Individual model results have been published over the past 30 years. The number of models has grown and the number of different publications increased. In 1995, the first **Coupled Model Inter-comparison Project (CMIP)**[1] was started to compare

---

[1]World Climate Research Program, "Coupled Model Intercomparison Project," http://cmip-pcmdi.llnl.gov.

© The Author(s) 2016

A. Gettelman and R.B. Rood, *Demystifying Climate Models*,

Earth Systems Data and Models 2, DOI 10.1007/978-3-662-48959-8_11

different models. The project has expanded in parallel with the **Intergovernmental Panel on Climate Change (IPCC)**[2] scientific assessments. Information and results from the current CMIP model submissions (the fifth round: CMIP5) were used in the 2013 IPCC Assessment Report,[3] the fifth such report. The CMIP project contains data from about 28 different modeling centers (groups of scientists that design and run climate models). The output from many different simulations for the past and future is generally freely available for use, and is being continually used in new ways by scientists and for applications research (as is described in more detail in Chap. 12). The results described in this chapter largely come from this large set of simulations. New models are in development for another round (CMIP6); the rounds occur about every 7 years or so. New models are typically released and run to coincide with the IPCC reports, at least over the past few cycles (from about 2001 to the present).

The IPCC scientific assessments are the best and most comprehensive entry point for looking at model predictions. In this book, we are concerned mostly with the physical climate system, which is treated by IPCC Working Group 1, whose focus is on the physical science basis for climate change. IPCC Working Group 2 looks at impacts, adaptation, and vulnerability of human systems; and IPCC Working Group 3 is focused on mitigation from an economic and policy perspective. Working Groups 2 and 3 also use physical climate models in their analysis. The reports are freely available from the IPCC. In the fifth IPCC assessment report, climate models are evaluated in Chap. 9, and results are spread throughout the IPCC report. Some selected results are reprinted in this chapter to illustrate the uncertainty in climate model results.

11.2   Prediction and Uncertainty

Before discussing results, it is worthwhile to frame the discussion of prediction and uncertainty by asking what we really want to know. This would seem like a simple question, but what we want to know, and how confident we are in the predictions, is very important for understanding how to use climate models.

Recall the terminology. We often use the word *prediction*, but we really mean *projection*. A prediction implies something will happen with some probability. A forecast is a prediction. There are multiple possible climate pathways, subject to several types of uncertainty. On long time and large space scales, the scenario uncertainty dominates (see Chap. 10). But only one solution will result. So we are really talking about multiple projections: if a given scenario happens, then this

---

[2]Intergovernmental Panel on Climate Change, http://www.ipcc.ch.

[3]IPCC. (2013). "Summary for Policymakers." In T. F. Stocker, et al., eds. *Climate Change 2013: The Physical Science Basis. Contribution of Working Group I to the Fifth Assessment Report of the Intergovernmental Panel on Climate Change.* Cambridge, UK: Cambridge University Press.

model *projects* that a certain climate will result. So the goal is to figure out (project) all possible states of the climate system. In some ways the multiple projections define a complete prediction of the possible states: This includes the scenario uncertainty and the model uncertainty. To really understand the possible future state, both types of uncertainty must be considered, but scenario uncertainty is currently outside of the models (it is an imposed forcing, generated with other models). Integrated assessment models would include this uncertainty and predict greenhouse gas emissions, but these models also have their own assumptions, like population projections that go into them.

Physical scientists like to think of a climate system forced by a human system. But the human system is really part of the climate system (the anthroposphere, discussed in Chap. 7). Thus projections of the human system are also possible, and are coupled with the climate system. Scenarios of human forcing for climate change also have constraints and can be modeled with human and economic system models (as opposed to physical models). These models also have to be realistic. That gets a bit more difficult. Whereas the laws of physics state that matter and energy is conserved, money is not conserved, and the "laws" of economics are complex rules that can change. But the structure of a society and an economy can be simulated, so that scenarios are also constrained to be realistic. For example, scenarios on emissions of greenhouse gases are usually based on projecting current rates of consumption and trying to estimate supply and demand curves for products like oil whose consumption will affect climate (by releasing CO_2 in the case of oil). The emissions then form the basis of future projections. Emissions scenarios are the predictions of integrated assessment models. The system is really a complex set of feedbacks because these models have their own projections input to them (like population).

11.2.1 Goals of Prediction

What exactly do we want to know from climate models? The basic answer is an estimate of the future climate, defined as the distribution of weather states. Climate is a function of space and time (summer climate, winter climate, particular location). We may also want an estimate of things derived from climate: stream flow, consecutive days above or below a threshold (heat and cold waves), or with and without precipitation (floods and droughts).

The climate of the future is a function of the uncertainties we have discussed: initial condition uncertainty, model uncertainty, and scenario uncertainty. On short timescales (a few days to a decade), initial condition (and internal variability) uncertainty dominates, and on century (or half-century) timescales, scenario uncertainty dominates. Model uncertainty remains over time, but may be significant. Model uncertainty remains significant at small scales, where model structural errors are important: If a model puts the storm track in the wrong place, it can consistently move storms (i.e., weather, precipitation) in incorrect ways and create an error in the climate at a particular location, even while the larger-scale climate (averaged over a continent, hemisphere, or the globe) is accurate.

The basic desire is to project all possible future weather states, given a set of forcings. Usually there are particular aspects to climate that we seek to understand, depending on our needs, and usually it is the extreme events (the low-probability tails of the distribution of states) that are of concern.

11.2.2   Uncertainty

What makes a prediction useful? What really makes it useful is a good estimate of the confidence (positive connotation) or uncertainty (negative connotation) in a prediction. In fact, a prediction without uncertainty may not be helpful. Is it going to rain tomorrow? The answer might be "more likely than not" (a greater than 50 % chance of rain). This forecast might be adequate if you are just going to work. You can take an umbrella along, just in case. If you are biking to work, you might need different clothes. But, say you are going to an outdoor wedding in the evening. That situation demands a more detailed answer: Will it rain now or later? In this case, the timing is important. The point is that the necessary confidence to make the forecast adequate depends on the situation. Even for one person, the needs of a prediction change with the purpose.

Another example might be to ask about changes in the availability of water in a particular place. Will it be wetter or drier in the future in my neighborhood? The timescale for "future" is important. How much will it rain next week? Next season? Next year? The next 30 years? "Next week" determines whether the garden should be watered. "Next season" may matter for a farmer. "Next year" may matter for someone managing a reservoir, and "the next 30 years" for someone building a reservoir.

Uncertainty is also a function of the spatial scales involved. Global estimates such as temperature or precipitation changes are more tightly constrained by the nature of the physical climate system, whereas regional changes on the continental or smaller scale are much more uncertain. One example is illustrated in Fig. 11.1, where uncertainty is broken into two dimensions.[4] One dimension involves the different pieces of evidence that are available, such as observations, models, and theory. Having more pieces of evidence is better. However, the other dimension is the consensus or agreement among those lines of evidence: If models, observations, and theory are available (three for three) and there is agreement, then results are "established" or have high confidence. However, if the three lines of evidence disagree, then there are competing explanations. Likewise, one or two lines of evidence may agree, but the agreement is incomplete. Speculation results from

---

[4]Moss, R. H. (2011). "Reducing Doubt About Uncertainty: Guidance for IPCC's Third Assessment." *Climatic Change, 108*(4): 641–658. doi:10.1007/s10584-011-0182-x.

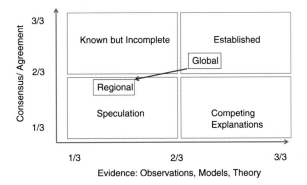

Fig. 11.1 The different dimensions of uncertainty. The *horizontal axis* is the number of pieces of evidence from observations, models and theory. The *vertical axis* is the consensus or agreement level. From Moss (2011), Fig. 2

having only one line of evidence available. Typically, it is easier to establish evidence at the large scale (global), and as we go to smaller scales the evidence and consensus may break down. This will clearly be seen in some of the figures that follow from model results. Smaller scales will show less agreement than global scales.

11.2.3 Why Models?

Why should we use climate models for prediction? Models are uncertain. They are likely to be wrong in many ways. The statistician George Box said "essentially, all models are wrong, but some are useful"[5]. Philosophically, all models are wrong in some way. The implication is that we can use an imperfect model if we understand its limitations. As we have shown, climate models are built on physical laws. These provide varying levels of certainty for models. Energy and mass need to be conserved. The fundamental laws of motion and many other physical processes are known, but they need to be approximated to match the scale of global models (tens of kilometers). Or, the processes are unknown or uncertain, and assumed. Note that different climate models may be more or less useful for different predictions, depending on how well they represent particular processes. That is part of the evaluation of climate models.

The evaluation proceeds from a simple philosophy. Because the laws of the physical climate system are based on physics and chemistry principles that are invariant, we can develop a model based on the present that can simulate the present

---

[5]Box, G. E. P., & Draper, N. R. (1987). *Empirical Model-Building and Response Surfaces*. New York: Wiley, p. 424.

and the past. This model can then be used to predict the future. The principle of weather forecasts is based on this: We evaluate past forecasts to better understand confidence in forecasts of the future. But weather forecasts can be evaluated every day. We do not have that many "climate states" to evaluate. We do have past climates (paleo-climate) as well as recent past climates. The goal of the evaluation of models is to determine what they are useful for, recognizing that they are imperfect representations of the climate system. More details of evaluation were discussed in Chap. 9.

11.3   What Is the Confidence in Predictions?

The key part of this chapter is assessing the confidence in predictions. We have already noted that having multiple lines of evidence agree is critical, and global estimates are more certain than those for smaller regions. Now we turn to specific examples and classes of model predictions to examine when models are established, or confident, to when they are speculative. For illustration, we use results from the most recent (2013) Intergovernmental Panel on Climate Change (IPCC) assessment.

The IPCC uses a controlled vocabulary for uncertainty, and links that vocabulary to quantitative statistical language.[6] One goal was to provide precision to terms such as *almost certain, unlikely,* and *doubtful.* Table 11.1 from the IPCC supporting material for the 5th assessment report[7] illustrates that the assessment language means something specific.

Table 11.1 Terms and likelihood estimates

Term[a]	Likelihood of the outcome
Virtually certain	99–100 % probability
Very likely	90–100 % probability
Likely	66–100 % probability
About as likely as not	33–66 % probability
Unlikely	0–33 % probability
Very unlikely	0–10 % probability
Exceptionally unlikely	0–1 % probability

[a]Additional terms (extremely likely: 95–100 % probability; more likely than not: >50–100 % probability; and extremely unlikely: 0–5 % probability) may also be used when appropriate

---

[6]Moss, R., & Schneider, S. H. (2000). "Uncertainties–Guidance Papers on the Cross Cutting Issues of the Third Assessment Report of the IPCC." *World Meteorological Organisation*: 33–51.

[7]Mastrandrea, M. D., Field, C. B., Stocker, T. F., Edenhofer, O., Ebi, K. L., Frame, D., et al. (2010). *Guidance Note for Lead Authors of the IPCC Fifth Assessment Report on Consistent Treatment of Uncertainties*. Intergovernmental Panel on Climate Change (IPCC), 2010. Retrieved from http://www.ipcc.ch/pdf/supporting-material/uncertainty-guidance-note.pdf.

11.3.1 Confident Predictions

We have confident predictions when we have multiple lines of evidence that indicate the same thing. These would be things that models are likely or very likely to be correct about. This would apply to one of the broadest metrics of a change in the climate: the global average surface temperature. But it also illustrates the complexities of prediction and some of the uncertainties.

11.3.1.1 Temperature

First, a definition: By surface temperature, we mean the temperature of the air near the surface of the earth. Practically, this is the temperature a weather forecast gives, and what you would feel if you walked outside. At any place and time it varies tremendously, but if you add up the temperature everywhere for a year, you are just measuring the heat content of the whole planet's surface, and this has to be constrained by the amount of energy in the system.

In Fig. 11.2, the scenarios indicate different increases in greenhouse gases from the Representative Concentration Pathways (RCPs) described in Chap. 10 and shown in Figs. 10.7 and 10.8. More greenhouse gases result in trapping more heat. We have to put the energy somewhere. The uncertainty in where it goes can alter the curves in Fig. 11.2. Heat can go into the ocean (especially the deep ocean). Heat can be reflected away by brighter surfaces. This uncertainty of variability in where the heat goes is what makes the temperature each year a bit different from the next and gives the "wiggles" on the curve. But many of those pathways still result in climate changes: If the surface temperature stays similar, the heat goes into the ocean and it may come out again, and it may change the ocean circulation. If the

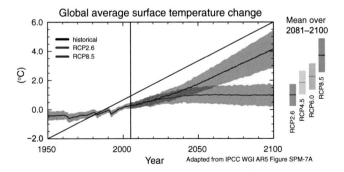

Fig. 11.2 Simulated global average surface temperature from climate models. Historical simulations in *gray*. Future RCP2.6 simulations in *blue*, RCP8.5 simulations in *red*. *Darker colored lines* are the multi-model mean. Also shown are the last-20-year mean and range from RCP2.6 (*blue*), RCP4.5 (*light blue*), RCP6.0 (*orange*), and RCP8.5 (*red*). Figure adapted from IPCC Working Group 1 Fifth Assessment Report, Summary for Policy Makers, Fig. 7a

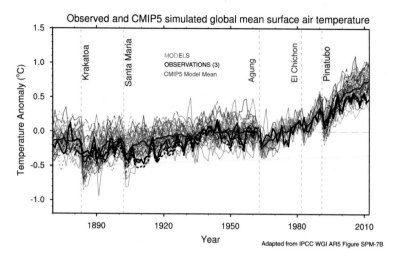

Fig. 11.3 Global average temperature anomalies, 1870–2010. Global average surface temperature anomalies from IPCC WGI 5th Assessment report models. Different models are different *thin colored lines*. The multi-model mean is the thick *red line*. Different temperature data sets are thick *black lines*. Anomalies taken from 1961 to 1990. *Dashed vertical lines* indicate the dates of named volcanic eruptions. Figure adapted from IPCC Working Group 1 Fifth assessment report, Fig. 9.8

clouds get brighter and cool some parts of the planet, some places get colder, while others can still warm. Even with no net global change, there can still be regional changes in climate.

The **spread** in results from the different formulations of climate models is due to different forcing and different responses to forcing. The response to a forcing includes all the feedbacks in the system that determine (in Fig. 11.2) how the system will respond to climate forcing. A larger response to the same forcing is a more sensitive climate (higher sensitivity). Note in particular that given the spread in models (model uncertainty, the shaded regions in Fig. 11.2), the scenario uncertainty is dominant after 2060 or so (the shaded regions no longer overlap). The model predictions of this scenario are within these bands, and thus the resulting surface temperature really depends on the scenario uncertainty after the middle of this century. The current generation of models is not likely to be very wrong. Counting on the scientific community to be very wrong on these metrics is not a prudent strategy.

The global surface temperature record is far from smooth, which makes understanding and attributing these large-scale curves difficult. In fact, there is plenty of evidence of the "fits and starts" to the global average surface temperature. This is illustrated in Fig. 11.3, showing the historical temperature changes from 1870 to 2010 from observations (black lines) and model simulations of these changes. There are multiple black thick lines reflecting different observational data sets: these overlap for the last 50 years or so but deviate from each other before that. Some of the big sudden dips in models and observations occur due to volcanic eruptions that put a lot of sulfur gas (sulfur dioxide, SO_2) into the stratosphere. The gas condenses rapidly to form small aerosol particles (not unlike water vapor condensing to form water or ice particles). These particles are smaller than cloud drops, but they do reflect light, and scatter some of it back to space, thus cooling the planet. The gas emissions are estimated and put into models (climate models do not "predict" volcanoes).

The other wiggles on a year-to-year timescale may be due to things like El Niño, which warms the planet a bit during the "warm" phase and cools a bit during the "cold" phase. The overall fits and starts in the trend (such as the cooling from 1950 to 1970, and warming from 1980 onwards) are due to the forcing: the combination of increases in greenhouse gases (warming) and increases in cloud brightening due to aerosols (cooling).

Also note in Fig. 11.3 the flattening of the trend, particularly in observations, over the past decade. The models generally do not see quite as much of a change in trend. The reasons for the reduced trend are unknown, but there has been potentially an increase in small volcanoes in the past decade, as well as possible changes to the ocean circulation. This is still an active area of research. The years 2014 and 2015 were the warmest on record, so this temporary hiatus in the rise of the global average temperature (similar to what happened from 1950 to 1970) may be fading.

So we are adding more energy to the system by trapping more heat with greenhouse gases. That heat has to go somewhere. Models project at least some of it is warming the planet. This record goes in fits and starts with the vagrancies of year-to-year variability, but models broadly agree with the historical record in Fig. 11.3. Model predictions of temperature changes are based on sound theory and the basic physics of the energy budget.

Could all models be wrong? This would require a very different understanding of the physics of the earth system. The magnitude of the warming is highly uncertain due to differences in treatment of the feedback response to a forcing, largely due to clouds. But it is very difficult to make a climate model *not* warm in response to adding greenhouse gases and still have a good representation of the present-day climate system. The upper bound on the range is less certain, but it is also hard to make a model too sensitive in the present-day climate. And the simple stability of the climate system over the past record (millions of years) is some evidence that the system cannot be "too unstable." So although it is possible for these projections to be wrong, it is unlikely.

One further aspect of changes to surface temperature is also observed: the tendency for high latitudes to warm more than the global average. This is seen in Fig. 11.4, based on model projections. This amplification of climate change at high latitudes is a consequence of strong regional feedbacks related to surface albedo: As the polar regions warm, snow and ice melts, and a strong positive local albedo feedback means that the now-darker land and open ocean surface can take up more heat. Also, because the heat in polar regions is transported from the tropics and then lost to space in the infrared wavelengths (see Fig. 5.3), more greenhouse gases interfere with the heat loss, and cause a larger warming. The tropics get heat from the sun and export it, and so the balance is not as affected by the infrared (long wavelength) changes due to greenhouse gases. The amplification is robust, but the

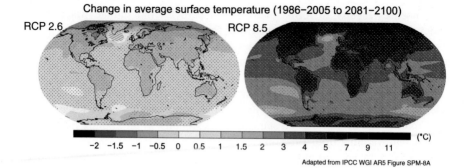

Fig. 11.4 Map of simulated average surface temperature change between the present (1986–2005) and the future (2081–2100) from IPCC WGI 5th Assessment report models (Moss and Schneider 2000). Stippled regions represent significant changes, hatched regions non-significant changes. Changes are larger in simulations using RCP8.5 (*right panel*) than RCP2.6 (*left panel*). Figure adapted from IPCC Working Group 1 Fifth assessment report Summary for Policy Makers Fig. 8a

quantitative magnitude of the amplification is not that certain (as it is not constrained).

11.3.1.2 Precipitation

Perhaps a more critical question when trying to understand predictions of climate change is to understand what will happen to precipitation. The amount and timing of precipitation is important for the climate of a place. As indicated in Chaps. 5–7, water is an important part of the energy budget of the climate system for the atmosphere, ocean, and land. Since we know the mass of water and the energy are constrained, this puts large-scale constraints on changes to precipitation. The constraints are strong on a global basis and get weaker when regional changes are considered. There are few large-scale constraints on the small scales of precipitation that we normally view as "weather" (perhaps the most important kind). Precipitation predictions fall into the "likely" category, with global estimates "very likely."

On the global scale, water in the atmosphere has to be evaporated from the surface of the earth, and this takes energy. So the amount of precipitation globally has to be balanced with the amount of evaporation from the surface. In a warmer world, the atmosphere holds more water vapor, by the simple physical law that warmer temperatures allow more water to remain in the vapor phase. The increase of water vapor is about 6 % per degree centigrade of warming. This value is seen in climate model projections, and it is also seen in observations of water vapor as the planet has warmed. So the air can hold more water, and there is more energy to evaporate water for precipitation.

There is no explicit law for how much precipitation will increase, but most model simulations indicate that the increase is about 2 % per degree centigrade of warming.

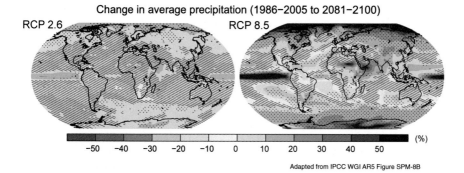

Fig. 11.5 Map of the model simulated change in average precipitation between the present (1986–2005) and the future (2081–2100) from IPCC WGI 5th Assessment report models (Moss and Schneider 2000). Stippled regions represent significant changes, hatched regions nonsignificant changes. Changes are larger for RCP8.5 (*right panel*) than RCP2.6 (*left panel*). Figure adapted from IPCC Working Group 1 Fifth Assessment Report, Summary for Policy Makers, Fig. 8b

Why is this the case, and why is it robust? Precipitation is controlled by the surface energy budget, and climate scientists often compare the mass of water in the atmosphere to the energy hitting the surface of the earth based on how much energy is used to evaporate water (latent energy). The amount of precipitation is a function of how much heat can be radiated away in the atmosphere to balance the latent heat of condensation. The increase in energy is not as fast as the increase in the available water vapor in the air. There is not a fixed law for this, but it is a result of many models.[8]

The enhanced water vapor in a warmer world is also expected to change precipitation patterns, as illustrated in Fig. 11.5. Since the air holds more water, in the tropics where the air is rising, more of this water is condensed, leading to increases in the average precipitation. The intensity of the **upwelling** needs to be balanced in the subtropical regions by **downwelling**, which is also projected to increase, drying these regions of the planet. The pattern can be seen in Fig. 11.5 from most of the models. The tropics get wetter, and the subtropics get drier. The changes to the general circulation are "fairly" robust (they are "likely")—based on sound arguments and models—but we do not have any proven theory or clear observations, so this only rests on one or two pieces of evidence (see Fig. 11.1) even at the global level.

When we think about specific regions and regional precipitation, then these large-scale arguments about global averages no longer are a constraint, and regional results are far less certain. The key to thinking about more certain predictions is understanding what they are based on, and whether the observational constraints are good (like the global energy being nearly in balance), whether there are good observations, and whether the models are effective at reproducing observations for

---

[8]He, J., Soden, B. J., & Kirtman, B. (2014). "The Robustness of the Atmospheric Circulation and Precipitation Response to Future Anthropogenic Surface Warming." *Geophysical Research Letters, 41*(7): 2614–2622.

the historical period. None of these is a guarantee, but it points to some sense of understanding, observations, and reproducibility that increases confidence in projections (see Fig. 11.1).

11.3.2 Uncertain Predictions: Where to Be Cautious

Precipitation is a classic case where some aspects of the impacts of climate change are well known, and some are much more uncertain. As noted briefly in Chap. 10, as the spatial scale of interest decreases, the large-scale constraints fall away, and potential model structural errors start to become larger.[9] While models agree on the sign and even some of the magnitude of global trends, they do not agree on the magnitude (even the magnitude of global changes), and particularly on what happens in different regions. These projections are less certain, or "as likely as not" in IPCC language from Table 11.1.

For temperature and precipitation, the broad regional patterns (wetter tropics, drier subtropics, warming high latitudes) are known, but the details of those patterns are highly uncertain, as is clear from Figs. 11.4 and 11.5. "Broad scale" means relying on the global energy budget, and global trends are fairly certain, but other classes of results are less certain.

In particular, the magnitude of many of these changes is not well known. While most models predict that the polar regions will warm faster than the rest of the planet (Fig. 11.4), the magnitude and speed of the warming is not well constrained. In addition, along with such warming as we have already seen in the Arctic has come a dramatic reduction in the sea-ice coverage (area).[10] This is illustrated in Fig. 11.6a from models. The region in gray is the spread of model simulations of the historical period. Models are pretty good at following the observed decline of the Arctic sea-ice extent, but they do not fully capture the magnitude (steepness) of the trend. Here is a case where projections indicate that in September the Arctic will likely be mostly free of ice by some date in the 21st century; it is mostly a question of when.

But that is not to say the models are doing that well. If we look in the Antarctic sea ice (Fig. 11.6b), which is generally more stable than the Arctic sea ice, models are predicting slight declines over the past 30 years, whereas the observations indicate increases in the extent of late summer sea ice. The spread of models is also very large. So while models seem to represent the Arctic well, it is not clear that they represent the Antarctic well. The reasons are complex and likely have to do

---

[9]Hawkins, E., & Sutton, R. (2009). "The Potential to Narrow Uncertainty in Regional Climate Prediction." *Bulletin of the American Meteorological Society, 90*(8): 1095–1107.

[10]Stroeve, J. C., Serreze, M. C., Holland, M. M., Kay, J. E., Malanik, J., & Barrett, A. P. (2012). "The Arctic's Rapidly Shrinking Sea Ice Cover: A Research Synthesis." *Climatic Change, 110*(3 4): 1005–1027.

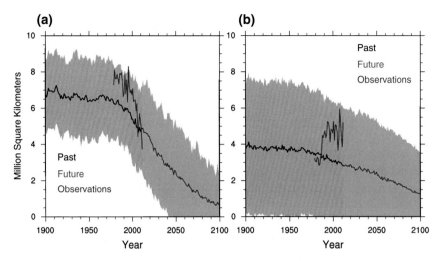

Fig. 11.6 Fall sea ice. Past and future simulated sea ice extent in the **a** Northern (September) and **b** Southern (March) Hemispheres. Model simulations for RCP8.5 in *black* (past) and *red* (future). Multi-model mean is the *thick line*. Observations of sea-ice extent for each hemisphere are shown in *blue*. Model simulations are from the 5th IPCC assessment report. Figure from Jan Sedlacek, ETH-Zürich

with interactions among the ice, ocean, and atmosphere. Our observations around Antarctica are spotty, and this may contribute to the lack of constraints on models.

Similar issues occur at smaller spatial and time scales. Although the high-latitude warming "amplification" seen in Fig. 11.4 is robust, the magnitude of the warming is widely different among climate models. This obviously is also true of the global average surface temperature change: For a given scenario in Fig. 11.2, the spread of estimates of surface temperature change by the end of the 21st century is nearly 2 ° C (5 °F), which is half of the 4 °C (8 °F) multi-model mean change. This is a large uncertainty. Obviously narrowing this uncertainty, and continuing to push models to better resolve smaller-scale features, is one of the goals of climate modeling and model development. We discuss extreme events that are highly uncertain in Sect. 11.3.4, after we discuss possible "bad" predictions.

One area where models do predict increases in extreme events is an increase in heavy precipitation. With more water vapor in the air, and a change to the cooling of the atmosphere, the regions of upward motion (which causes rain) are expected to increase their vertical motion, and perhaps decrease their extent. The increasing vertical velocity and more water vapor in the air would drive increased moisture convergence at low levels, and more intense precipitation. The magnitude is uncertain, and the mechanism is also somewhat uncertain, but most models show such an effect with warming of the surface.

Another uncertain prediction concerns the role of the carbon cycle in the future. As discussed in Chap. 7, currently the ocean and land surface take up about half of the CO_2 emitted by humans. In terrestrial ecosystems, increasing concentrations of

CO_2 may yield higher growth efficiency. The growth of plants pulls carbon dioxide from the atmosphere into plant tissue and into the soil. Thus the terrestrial carbon uptake is more efficient and may increase CO_2 uptake. Many climate models that include a carbon cycle predict this effect. But it is uncertain because there are competing effects: Plants may grow more efficiently and use water more efficiently with higher CO_2, but increased heat stress may reduce growth. Because the result comes from a balance of offsetting uncertain processes, that makes the net effect uncertain. It is also an effect easy to observe in a controlled experiment, but it is hard to scale up such observations to a global-scale carbon uptake.

So when are projections likely to be uncertain? When there is less of a constraint from the physics of climate or observations. For example, the changes to water vapor are fairly certain, because they are based on proven physical laws. The changes to precipitation are a further step removed from those physical laws because they involve more complex cloud processes, and they are therefore less certain. The changes to the general circulation are fairly certain, but specific regional changes are less certain. The changes to the carbon cycle rely on compensating effects, which are probably even more uncertain.

11.3.3   Bad Predictions

Climate models are not perfect, and they are only as good as the observations and our understanding. Where observations and understanding are lacking and uncertain, we are in the space of Fig. 11.1. Where we only are looking at a weak pillar of knowledge (closer to the origin in Fig. 11.1), then predictions based on that understanding will also be highly uncertain, and they may be totally wrong. Understanding where models are likely to be wrong, or where they are likely to have the range of projected impacts change (expanded or moved), is critical for assessment of model results.

What do we mean by a "bad prediction"? Generally, a bad projection would be where the actual result is outside of the error bars or uncertainty range that we specify for a particular parameter or metric. The result is "unexpected." These are places to watch out for. Bad predictions usually result from not understanding the uncertainty, and making predictions based on models that are uncertain or are not well backed up by observations and theory. This is also called Overconfidence. In this context, a projection based on model output becomes "bad" if the uncertainty is wrong. In Fig. 11.2, if the spread in models for the future looked like the past (very small spread), then the odds of the projection's being wrong would be much higher. So one of the best ways to avoid bad predictions is to be very careful about understanding total uncertainty.

In general, such lack of understanding of uncertainty (which is often not properly expressed) comes from uncertainty in knowledge (theory) or in observations. Perhaps the best example of this is the projection of sea-level rise due to climate change, where the range of estimates from models (and expert judgment) continues to change,[11] and where models have a hard time simulating the sparse observations available. Sea-level rise occurs because of melting of land-based glaciers and ice sheets that adds water to the ocean. But it also occurs because increasing the temperature of the ocean causes it to expand and take up more volume (thermal expansion). In addition, there are local changes in the land surface due to rebound after melting of ice sheets from the last ice age. Parts of North America are rising or sinking relative to the ocean because the tectonic plates of the earth's crust are still adjusting to the removal of the ice sheets from the last ice age, in a way similar to how a piece of soft foam gradually restores it shape after a weight is removed. Generally we only think of the first issue: adding water to the ocean. Melting floating sea ice does not change the sea level, in the same way that ice melting in a glass does not cause the glass to overflow.

Predictions of sea-level rise are changing as we learn more about ice sheets. In particular, the Greenland ice sheet is thought to be critical for global sea-level rise due to ice sheet melting. Sea level rise projections made in 2013 explicitly stated that they were estimating the ocean thermal expansion only, and could not quantitatively estimate the contribution of additional ice sheet melting to sea level rise.[12] Thus, taking the model projection as being representative ignores the uncertainty and may underestimate the change.[13] Current projections now try to simulate the ice sheets themselves, and to take into account the dynamics of ice sheets, particularly Greenland. But models have a hard time reproducing the present estimated rate of loss of the ice sheet. This current loss is occurring because of melting and because of changes to the flow of the ice sheet. Constraining the mass of the Greenland ice sheet is difficult. Estimates of the extent and elevation (volume)[14] are matched with satellite estimates of the mass of the Greenland ice sheet.[15] But both estimates have uncertainties much larger than the estimated mass loss, so that is not much of a constraint. Estimates are also made from regions where the temperature is above freezing and melting is occurring.[16] But all this adds up to lots of uncertainty.

Recently, additional processes have been discovered that can change the flow of the ice sheet, such as water flowing down from the surface through cracks in the ice sheet to the base, where it potentially can make the base of the ice sheet easier to

---

[11]See, for example, Bamber, J. L., & Aspinall, W. P. (2013). "An Expert Judgement Assessment of Future Sea Level Rise From the Ice Sheets." *Nature Climate Change, 3*:424–427.

[12]These projections of sea-level rise come from the 2013 IPCC report.

[13]Rahmstorf, S., Foster, G., & Cazenave, A. (2012). "Comparing Climate Projections to Observations up to 2011." *Environmental Research Letter, 7*:044035.

[14]Zwally, H. J., Giovinetto, M. B., Li, J., Cornejo, H. G., Beckley, M. A., Brenner, A. C., et al. (2005). "Mass Changes of the Greenland and Antarctic Ice Sheets and Shelves and Contributions to Sea-Level Rise: 1992–2002." *Journal of Glaciology, 51*(175): 509–527.

[15]Veliconga, I., & Wahr, J. (2005). "Greenland Mass Balance From GRACE." *Geophysical Research Letter, 32*:L18505. doi:10.1029/2005GL023955.

[16]Van den Broeke, M., Bamber, J., Ettema, J., Rignot, E., Schrama, E., Jan van de Berg, W., et al. (2009). "Partitioning Recent Greenland Mass Loss." *Science, 326*(5955): 984–986.

slide.[17] Enhanced flow can make glaciers at the edge of ice sheets flow faster (more icebergs). Models are attempting to simulate this. But the current versions of ice sheet models, even when trying to simulate these bottom (basal) lubrication processes, have not been able to get much mass loss at a rapid rate, and not as fast as observations over the past 20 years or so. This is a serious deficiency in model evaluation, and one reason why projections of sea-level change are so uncertain. But as new processes are discovered, this may change. Or maybe estimates will be revised downward as we better understand the simulations and pieces of them. As long as we know what is missing and what the uncertainty is, we can gauge whether a prediction is wrong, and also in what direction. Is a projection an overestimate or underestimate? Or an upper or lower limit?

Thus, what really makes a projection "bad" is overconfidence, or underrepresentation of uncertainty. Often uncertainty is stated somewhere, but not presented well or ignored. The lesson is always to try to understand a projection's stated uncertainty. This is true in general, not just for climate models. The best practice for using models is to go back to the model documentation or description to make sure a proper representation of uncertainty is available, and an analysis of the model fit for the purpose is assessed. For example, projections of changes to a phenomena based on models with a bad representation of the present phenomena (like the South Asian summer monsoon, for example) may fall into this category.

11.3.4 How Do We Predict Extreme Events?

Some special mention needs to be made for extreme events. These are the infrequent tails or extremes of a distribution, which occur with low probability. No one gets killed by the global average temperature: It is extreme events at local scales that cause havoc and damage. How do models simulate these events? What category do they fall into? Many types of extreme events are well observed and predicted from hours to a week in advance by weather models.

There are a diversity of extreme events at the tails of the distributions that make up climate. We can easily envisage high and low temperature extremes, and damaging rainstorms. Tropical cyclones (hurricanes), windstorms, and snowstorms are further examples. But extreme events also happen in time. One day of record-high or -cold temperatures is one thing, but a sequence of events together like a heat wave, where temperatures remain high, is even more damaging. Or take a month with above-average rainfall at a location. If it occurs evenly over the month, perhaps that is not so bad, but if all of the rain occurs in three consecutive days, that could be a real problem.

---

[17]For an overview and some great pictures, see Appenzeller, T. (2007). "The Big Thaw." *National Geographic, 211*(6): 56–71.

One reason that extreme events are hard to project is that they are often hard to simulate in the present day. There are several classes of events that fall into this category. **Persistent** events, such as droughts or heat/cold waves are good examples. They are large-scale, but infrequent, and may depend on complex interactions that give rise to stationary patterns. Even weather models at fine scales often have a hard time predicting these events.

Another reason why extremes are hard to project is that they occur on small spatial and time scales. Extreme precipitation events and floods generally occur in local regions, based on local topography of a single valley that cannot be represented in a global model. Or the interactions may be small scale and depend on resolving small-scale features such as tropical cyclones, which have known, but complex circulations (e.g., swirling rain and cloud bands, a dry "eye" at the center).

So what can climate models say and how? As discussed earlier, there are often two ways of projecting extreme events. One is to try directly to simulate them, which for persistent heat and cold, or dry and wet events, should be possible in climate models. As yet, models are struggling with representing the stationary flow patterns observed in the atmosphere that give rise to many of these events. These persistent patterns are blocking patterns, mentioned in Chap. 10: a different than usual flow of weather systems that persists for a few weeks due to a stationary pattern in the large-scale storm track. Blocking events can steer storms in particular ways. The impacts of El Niño on western North America result from the tendency of the tropical Pacific temperatures to affect the position of the storm track hitting the west coast. During warm events, the storm track makes landfall in the south, bringing wet conditions to southern California, but keeping the Pacific Northwest dry; the opposite occurs during cold events. These large-scale effects can be simulated directly, and large-scale persistent events should be able to be simulated.

The other way to simulate extremes and how they might change is to use proxies (see Chap. 10) or downscaling the models (see Chap. 5). This is often done by looking at the large scale and developing a physical or statistical relationship between the large scale and the extreme events. We illustrate a few application examples in the next section.

11.4   Climate Impacts and Extremes

There are many different dimensions of using climate models to estimate impacts of climate change, typically by estimating changes to extreme events. Here we briefly present a few examples: first, the application of climate models to predict tropical cyclones, and, second, the application to provide a future distribution of stream or river flow. Both of these methods typically involve downscaling predictions in various ways. Stream flow typically also involves coupling to a physical model of a watershed and stream. Finally, we look at using climate model output to simulate electricity demand, which focuses on temperature extremes. Applying climate models to human systems requires coupling to a model of energy use and demand: a partial model of the anthroposphere.

11.4.1   Tropical Cyclones

Tropical cyclones are an important and relatively small-scale atmospheric phenomena. A climate model at low resolution will not adequately represent tropical cyclones, but it will typically have weak versions of them: warm-core cyclonic systems in the tropics that propagate like tropical cyclones but with very low wind speeds. These can be estimated, and how these "pseudo storms" change in the future can be used as a guide. Or the "potential" for storms can be derived. For tropical cyclones, this is often based on an index derived from present conditions that predict average storm intensity from the large-scale moisture and wind fields. These indexes of "potential storms" or "potential intensity" can be estimated in climate models now and in the future. This is one example of downscaling discussed in Chap. 5: using large-scale output to represent what fine-scale structures should be present.

   The danger with a lot of these proxy or downscaling methods is the danger of overfitting to the present day: If a measure of tropical cyclones is based on sea surface temperatures of the tropical oceans, and the current maximum is 82 °F (28 ° C), how will this work if the maximum rises to 84 °F (29 °C)? We are out of the range of observations. There is no guarantee that the proxy based on sea surface temperatures will represent the same variability in cyclones now or in the future if we are forced to extrapolate a statistical model to future conditions that have not been experienced in the past.

11.4.2   Stream Flow and Extreme Events

We have been speaking of physical models, but derived impacts can also be coupled to climate model output in this way. These can range from physical application models to economic models. An example of a physical application model might be a model of stream flow in a particular watershed, based either on precipitation, or perhaps on precipitation and the wind direction (indicating where storms are coming from, and which slopes might receive their water). The inputs to the model might be precipitation and wind, and the outputs stream flow at a point on a river. Likely there would also be some downscaling involved to generate precipitation and temperature estimates for particular points or over a particular region that is not the same as a large-scale model grid. The model of stream flow could be a physical model related to precipitation and slope of terrain, conditional on the soil moisture (like a simple bucket model of runoff described in Chap. 7). But the stream flow might also be purely statistical, or empirical. If you take the historical observations based on a series of rain gauges: if there was X mm (in.) of precipitation in 24 h, then the stream flow was Y. As long as the future rain was never greater than X, you could estimate stream flow with a mathematical relationship (a regression) between the observed rain and stream flow.

11.4.3 Electricity Demand and Extreme Events

A more economic application might be the use of electricity (electricity demand) as a function of temperature in a particular region based on current patterns. Such a model could again be based on a model of the energy system, but would likely need to have an empirical component. For example, in the past, when the temperature was W degrees, then the electricity use was Z megawatts. It would be based on the current energy system.

In the example of electricity demand, it should be obvious by now that there are several dimensions of uncertainty. If the future temperature is out of range of current temperature, then electricity demand must be extrapolated. And the farther into the future you go, the less valid a statistical model based on the current energy demand as the system changes. Carefully identifying these uncertainties and assumptions is the key to prediction of extreme events.

The prediction of extreme events, whether directly or by indirect methods (proxy or downscaling) is quite difficult. It requires that reasonable assumptions be made about how events may or may not change in the future, and the best metrics for them. Direct simulation of many events may be possible, especially for large-scale persistence events (heat and cold waves), but downscaling methods will remain important. The key to using statistical downscaling is to limit extrapolating or overfitting. It is also critical to ensure that the model represents the base state well.

11.5 Application: Climate Model Impacts in Colorado

This case study demonstrates the direct use of temperature and precipitation data from climate model projections. Aspen, Colorado—a city in the Rocky Mountains —is known as a summer and winter recreation center. High value is given to the environment, and the political and economic environment supports a proactive approach to climate change. There are locally funded efforts to directly apply climate projections to city and regional concerns. The most visible issue being addressed is the future and the viability of local ski resorts. However, planners are also concerned about flooding and the potential for catastrophic mudslides.[18]

---

[18]Climate Change and Aspen Impact Assessment, 2014, http://www.aspenpitkin.com/Portals/0/docs/City/GreenInitiatives/Canary/GI_canary_ClimateChangeAspen2014.pdf.

The City of Aspen's area is approximately 5 mi^2 (12 km^2). The surrounding county is approximately 1,000 mi^2 (2,500 km^2), which is approximately the area of a 32 × 32 mile (50 km × 50 km) grid cell. Length scales for resolved weather features in a model with that grid size would be approximately 320 miles (500 km) on a side.[19] The area has steep topography, which strongly influences precipitation and the partitioning of water into watersheds. The topography is coarsely represented in the climate model. The native model information is, therefore, on a spatial scale that is far too large to apply directly to the city and county. Localization of coarser global climate model information by downscaling (see Chap. 8) can provide additional guidance for expert interpretation; however, it does not overcome the shortcomings of the global simulation or reduce uncertainty.

Application of model-generated data first requires evaluation on the spatial scales of interest, which brings attention to how well the model has performed over an observed time period in the past. Compared to a locale with smooth topography, there are larger uncertainties in the observations, especially for precipitation. The comparison with the past establishes the credibility of the model performance and contributes to the description of uncertainty. Straightforward comparison demonstrates that the temperature from the model compares better than precipitation, a nearly universal characteristic of climate models.[20] Precipitation has large errors relative to observations. The spatial structure of model and observed precipitation are poorly correlated at an individual grid point or even small clusters of grid points (3 × 3 or 5 × 5 grid points).

Standard practice in such applications is to look at the variability of an ensemble of climate models (see Chap. 10). This, potentially, reveals models that compare better to observations in a local region. This is also one of the more robust measures of uncertainty, specifically, a measure of model uncertainty. It is also a way to gain knowledge on the ability of models to span observed variability of, for example, extreme rainfall events.

Model biases at a particular place can often be traced back to specific processes. For example, summertime and wintertime precipitation processes differ at Aspen. There are two reasons for the seasonal difference. First, as in many land regions, summertime precipitation is caused by thunderstorms, represented in climate models by convective parameterizations. Wintertime precipitation is larger in spatial scale; however, wintertime large-scale precipitation is highly sensitive to topographical details. The intrinsic model error characteristics associated with convective and large-scale precipitation are different; hence, the error characteristics of summer and winter precipitation may be different.

Second, summertime precipitation in Aspen is associated with the North American monsoon, a regional monsoonal flow that brings moisture into the region

---

[19]Kent, J., Jablonowski, C., Whitehead, J. P., & Rood, R. B. (2014). "Determining the Effective Resolution of Advection Schemes. Part II: Numerical Testing." *Journal of Computational Physics,* 278: 497–508.

[20]Climate Change and Aspen Impact Assessment, 2006, http://www.aspenpitkin.com/Portals/0/docs/City/GreenInitiatives/Canary/2006_CCA.pdf.

from both Pacific and Gulf of Mexico sources. Wintertime precipitation is more often associated with large-scale weather systems with a history of propagation over the Pacific Ocean and crossing the span of mountains between the Pacific coast of the United States and Aspen. Thus, the moisture sources and relationship to global climate processes (e.g., El Niño) differ seasonally, and the biases may be different in different seasons.

Temperature is usually better represented than precipitation. Model performance and process analysis of precipitation reveal fundamental shortcomings. These shortcomings are not convincingly reduced by use of localization techniques such as downscaling. Using multiple simulations in an ensemble can aid in interpreting uncertainty; however, the different simulations in the ensemble may not reveal a class or subgroup of models that can be confidently chosen as best for the analysis.

Model guidance for planning follows by looking at time variability over the region across the ensemble. Averages (spatial, temporal, and ensemble) can be used to reduce random errors and quantify bias. Credibility and salience (relevance; discussed in Chap. 12) of model output are established by analyzing past trends and variability. If past trends and variability simulated by a model are established over a time span of several decades, then changes of behavior in projections of the future are imbued with credibility.

Of special note in this case study, the effort has been under way a number of years, crossing two versions of climate model experiments, from 2007[21] to 2013.[22] The 2007 simulations suggest a likelihood of warming with less precipitation. The 2013 simulations suggest more possibility of warming with more precipitation.[23] The uncertainty in precipitation is reflective of the challenges of calculating moisture transport to a region and conversion of this moisture to precipitation. This type of uncertainty can be managed in planning by consideration of plausible scenarios and decision making within those scenarios. This is followed by revisiting the projections as models improve and observations confirm or refute model behavior.

11.6 Summary

Good predictions have consistency among theory, observations, and models. Observations are a key part of having confidence in predictions. Bad predictions are often made because the uncertainty of an estimate is not known or is improperly

---

[21]Coupled Model Intercomparison Project, Round 3. Reported on in Solomon, S., Qin, D., Manning, M., Chen, Z., Marquis, M., Averyt, K. B., et al., eds. (2007). *Climate Change 2007: The Physical Science Basis: Working Group I Contribution to the Fourth Assessment Report of the IPCC*. Cambridge, UK: Cambridge University Press.

[22]Coupled Model Intercomparison Project, Round 5. Reported on in IPCC (2013). See note 3.

[23]See also Colorado Climate Change Vulnerability Study, 2015, http://wwa.colorado.edu/climate/co2015vulnerability/co_vulnerability_report_2015_final.pdf.

presented or translated. The way to avoid bad projections is to understand how the projection is built, and how certain it is likely to be. Bad projections are likely to result from models that are not being used for the right purpose.

With respect to current climate model simulations, we are likely to see warming, and model spread (uncertainty between models) is smaller than the uncertainty (difference) between possible emission scenarios. Thus, scenario uncertainty, not physical model uncertainty, dominates the global-scale prediction uncertainty.

We have higher confidence in temperature prediction than precipitation prediction. We have some confidence in global changes to the general circulation, but regional effects and magnitudes are highly uncertain.

Sea-ice predictions are uncertain: Models can do the right thing in the Northern Hemisphere, but they do not see the same trends as observed in the Southern Hemisphere. Sea-level rise projections are still uncertain as new processes are being added for ice sheet models, and these models currently have a hard time reproducing observations.

Perhaps one way of describing the goal for better climate prediction and improved models is to move more of the prediction uncertainties from the "unknown" category into the "more certain" category (see Fig. 11.1). The critical uncertainties are many that have been listed above. These include changes in regional patterns of precipitation, and changes in extremes of precipitation as well as temperature. Predicting the future of these events means representing the events well in the present-day climate, and being able to compare to detailed observations of extreme events—knowledge of the tails of the observed distribution of climate variables.

Key Points

- Climate models provide projections for the future but are dependent on scenarios.
- Scenarios are uncertain. Scenario uncertainty may dominate on century timescales.
- Global average temperature projections from models, and even regional projections of long-term temperature change, are well constrained.
- Precipitation changes are less well constrained in models.
- Projections of sea-ice extent and sea-level rise are highly uncertain.

Chapter 12
Usability of Climate Model Projections by Practitioners

Ultimately, a goal of climate modeling is to provide useful projections of future climate for policy makers and for "practitioners," those who need to make planning and management decisions based on climate. Practitioners include engineers, water resource managers, and urban planners.

The challenges of communication and use of model projections in planning and management is not trivial. The complexity of models is one barrier: We have used many words to describe the concepts in a coupled climate model. The complexity also comes from the basic difficulty in connecting causes to effects. Causes are emissions and concentrations of greenhouse gases, which first affect the heat input to the climate system (radiative forcing). Effects are climate impacts through the different parts of the system. There is the difficulty and uncertainty in connecting the forcing to a wide range of projected average temperatures and then to an even wider range of regional effects that vary from model to model.[1]

In addition to complexity in cause and effect, there is a more fundamental issue. Climate science relies heavily on simulation models. The use of simulation models often seems strange not just to nonscientists, but even for scientists trained in observational methods who focus more on statistics than on the underlying equations of a system. Therefore, there is a need to communicate the logic of modeling, which requires facing apparent contradictions. For example, one of the major contradictions we have attempted to address is that as models are made more complete, there is little reduction in the *headline uncertainty* (the uncertainty in the global average temperature change).

In this book, we have tried to provide engaged model users with an improved understanding of the logic of modeling, models, and their use in climate science. We have also described model performance and identified essential uncertainties. Even with this knowledge, the use of climate projections in practice remains difficult. There is a growing literature on the use of science-based knowledge, which in the case of climate science is partially motivated by the fact that despite the predictions of dangerous and disruptive climate change, there is relatively little real action. This chapter explores the use of model information, both conceptually and

---

[1]Pidgeon, N., & Fischhoff, B. (2011). The role of social and decision sciences in communicating uncertain climate risks. *Nature Climate Change, 1*(1): 35–41.

© The Author(s) 2016
A. Gettelman and R.B. Rood, *Demystifying Climate Models*,
Earth Systems Data and Models 2, DOI 10.1007/978-3-662-48959-8_12

with case studies. Our goal is to examine the processes involved in the use of model information so that we can help the reader overcome barriers to use of climate model output for improving policy and decision making.

12.1 Knowledge Systems

The literature on the usability of predictive geophysical models (like climate models) relies largely on case studies of the successful use of information. The use of weather forecasts in decision making is so common as to be intuitive. Forecasts of impending extreme weather (within 1–5 days) are used to plan emergency responses. Evacuations are called, or transport (like commercial airplane flights, or train service) is rerouted or cancelled. Other things happen as well, with less media attention. For example, when extreme weather is forecast for winter storms, snowplow drivers are asked to work overtime. When ice storms are predicted, utility crews are brought in from other states to be ready for downed power lines. Still, how to express weather-forecast information, risks and opportunities associated with weather forecasts, and the uptake of that information by decision makers, including individuals, is a subject of controversy and active research.

Perhaps more relevant to the usability of climate information are the studies of seasonal forecasts in decision making. To understand how climate model information is used, it is valuable to understand how "climate knowledge" (e.g., climate projections) relates to other forms of knowledge that are needed to address a particular problem. For example, to manage an ecosystem such as a wetland, climate model information on local precipitation and runoff might be applied as input into a model of the flow of water through the wetland, and the resulting water level. Decisions might need to be made regarding how much water flow is necessary to allow the wetland to function. The water flow may be regulated by an upstream dam, so that the water flow can be adjusted. The climate and derived ecosystem information then informs a portfolio of management possibilities that might be constrained by policy, politics, budgets, and the like. These management needs may include balancing the need to maintain a water level in the wetland, with the needs of water users for agriculture and the need to store water for another season, or to provide for flood control.

The important point is that the climate projection is only a part, and usually an input part, of the decision-making process. The climate information must be relevant to the decision-making process in order to be useful. This has important implications. First are simple things, such as having the right output data from the model (stream flow, or runoff or just precipitation) in the right units. Second are more complex aspects, such as understanding what the uncertainty on the forecast might be and how to reflect that in another application.

A near universal conclusion from the research on usability of information is that it is simply not adequate for climate projections (whether seasonal or longer term) to be placed into a data portal (i.e., made widely available) with the expectation that the projections can be broadly used by practitioners. For the most part, model projections are created by climate modeling groups that produce data, and then leave the data to sit in a metaphorical loading dock or shop window. These data may not ever be accessed. Successful use of climate data in decision making follows from an iterative human process with multiple directions of communication. Models are used by decision makers, and their questions and analysis are used to improve models. This back-and-forth process establishes the relevance of the climate information in the context of the problem.

Simulation output needs to correspond directly to critical inputs. It does not help if a model supplies daily averaged precipitation over a model grid box of ten thousand square miles (100 × 100 miles) if what is needed is hourly stream flow or hourly total runoff for a particular region like a city or a drainage basin, which may span parts of several model grid boxes. The back and forth iteration needs to be between a user (decision maker) and someone who can help interpret the model output (as we describe in the next section). Interpreters need to understand the appropriate and inappropriate use of climate model information. They also need to be experienced users (but need not necessarily build or run climate models themselves).

Putting forecasts into the process of decision making also needs to put the uncertainty associated with the climate projection in the context of the problem. How much of the uncertainty comes from uncertain climate information? The uncertainty discussion is often simplifying, with the realization that the uncertainty of the climate model does not have to be quantified in an absolute sense. This is especially true if the uncertainty associated with climate change is small relative to uncertainties associated with policy, engineered systems, and other attributes of the natural and built environment. As an example, the future impacts of tropical cyclones on a particular stretch of coastline may depend more on what buildings get built on the land than on changes to tropical cyclones impacting the location. An increase in hurricane intensity (wind speed) or frequency may change the expected loss by 50 %. But changing the zoning (what buildings can be built on the land) might result in going from low-density houses to a hotel: If ten $200,000 houses ($2 million) become a $10 million hotel, then because the value increases by a factor of five, the expected loss would increase as well by a factor of five (500 %).

The usability of climate data and knowledge by practitioners is often stated to depend on three things: legitimacy, credibility, and salience.[2] **Legitimacy** describes whether the forecaster is objective, fair, and free of other biases. Is the person making the forecast "legit?" **Credibility** in this context refers to whether the forecast is scientifically valid or credible. Together, legitimacy and credibility suggest the need for decision makers to establish trust in the information they are using: through both trust in the information provider, as well as trust in the model

---

[2]Cash, D., Clark, W. C., Alcock, F., Dickson, N. M., Eckley, N., & Jäger, J. (2000). *Salience, Credibility, Legitimacy and Boundaries: Linking Research, Assessment and Decision Making.* KSG Working Papers Series RWP02-046, http://ssrn.com/abstract=372280 or http://dx.doi.org/10.2139/ssrn.372280.

used. Most of the discussion in this book so far has concerned the scientific credibility of climate models. **Salience** requires that the information be relevant to a practitioner's problem. More than legitimacy and credibility, this chapter is concerned with salience or relevance, which is difficult to establish.

Relevance or salience often brings forward the need for the evaluation of the data from climate model projections. This is evaluation beyond that performed in modeling centers and through scientific research papers. The characteristics of this evaluation are, often, that it is highly local, is application specific, and uses different variables (derived indices) than provided or evaluated by modeling centers. An example might be taking temperature or maximum temperature and estimating heat waves, or the stream flow in a particular river. The evaluation requires linking past performance with interpretations of the future. Just because a model reproduces the global average temperature or precipitation, that does not mean that the model reproduces the important characteristics of precipitation (frequency and intensity) at a particular location. Further evaluation is often necessary to evaluate models as fit for a particular purpose. This evaluation step in the application of model data is necessary enough that model-data providers should conceive their data as the start of further evaluation rather than just focusing on the practitioner's direct application of the data.

12.2   Interpretation and Translation

The need to make climate model projections relevant to a specific application can be described as the need to translate the information and derived knowledge in the context of a particular problem. The 2012 report from the National Research Council, *A National Strategy for Advancing Climate Modeling*,[3] called for the development of a profession of **climate interpreters**. This recognizes the need to help establish salience. If salience needs to be established for each problem brought forth by practitioners, this represents an enormous task. Therefore, it is reasonable to expect that salience might be obtained in particular sectors or discipline areas of sectors (e.g., agriculture, water management, ecosystems, public health) or for regions with similar geography or climate. In these cases, groups of practitioners may be able to share basic information, for example, how freezing rain will change in the eastern half of North America.

Interpretation of climate model data is part of the necessary iterative process of the use of climate information. It is not simply recasting data and knowledge into a different form. Equally important is for the climate scientist or interpreter to understand the language and context of the decision maker. A simple example is given with the words *anomaly* and *trend*, from statistics, which is often the

---

[3]National Research Council. (2012). *A National Strategy for Advancing Climate Modeling*. Washington, DC: National Academies Press.

quantitative language that bridges fields. Words such as *anomaly* and *trend* may take on quite different meanings within the context of a specific field or application. For example, in a general context, an anomaly is something that does not fit (e.g., "the model is anomalous"), whereas in a climate context, the anomalies are usually a reference to a dataset with the average removed. In general, an anomaly in common use is bad, but in a climate timeseries, anomalous events may be a valuable part of the signal, and the word is neutral. *Bias* is another word that has a negative connotation in general usage, but in a climate science context, *bias* is another word for systematic difference from an observation.

More complex questions of interpretation are related to climate parameters that may act in counterintuitive ways, such as the likelihood that in a warming climate there will be more snow in individual storms, yet less snow cover in the late winter and early spring. Explanations and usability of such correlated behavior is not well served by simple metaphors. Likewise, robust and cogent explanations of such concepts are of little use if they are not easily found and included in a practice that connects generation of predictions (or projections) and applications.

Interpretive or translational knowledge exists within an environment that includes multiple paths of communication of knowledge and positions of many stakeholders. So interpretive or translational information is often needed in the application of climate model projections. Individually, this information is not sufficient to ensure usability; however, detailed information about model forecasts or projections is often necessary in the context of a specific problem.

12.2.1   Barriers to the Use of Climate Model Projections

An often-cited barrier to using climate model projections is basic information that describes the model and output products. A climate model needs a manual. The information to be documented includes, for example, glossaries that describe variables and file names, description of specific model configurations and experiment design, and underlying technical documentation of the equations used in a model. This information is used to inform quality, reliability, and trustworthiness (part of "credibility and legitimacy"). There has been significant effort in the climate community to provide this basic information; however, its usability often requires discipline expertise from members of the climate science community. Hence, even at this initial phase of application there is a need for interpretation and translation.

Beyond this basic information, another frequent barrier to the usability of climate data is the fact that the standard data format for climate models is not familiar outside of the climate community. Furthermore, the data format standards for observational datasets differ from those of model datasets. Likewise, standards and conventions of gridded and ungridded datasets differ. An example is the simple difference between gridded data (on a regular horizontal grid) and station locations measuring stream flow along a river. Formats such as those associated with geographic information systems (GIS) are far more common in the practitioner communities. GIS uses a common format for referencing data to a spatial grid on the earth (geospatial), and is used in many planning and engineering fields requiring geo-location (e.g., flood control, city planning). To facilitate exchange, there now exist archives that have "translated" climate model output in common GIS formats. In addition, the practitioner's applications often have far more spatial information than climate models, which can challenge the salience of the climate model data. For example, there are also important differences in how climate models and practitioners' tools represent the quasi-sphericity of the earth.

12.2.2   Downscaled Datasets

A strategy commonly used to address salience is spatial downscaling (see Chap. 8). The spatial resolution of climate models (tens to hundreds of miles or kilometers) is coarse compared to the spatial scale desired by many practitioners, which may be as small as parks within cities, or similar scales to capture the irregular boundaries of watersheds or catchment areas. Many spatial downscaling techniques have emerged, and many practitioners go directly to these processed and downscaled datasets, rather than the original simulations, for their applications. In many cases, these spatially downscaled datasets have also had local bias correction to align the simulated means from the models with historical observations. For example, if the mean temperature of a place is 68 °F (20 °C) and a climate model indicates 70 °F (21 °C), then 2 °F (1 °C) are subtracted from each "simulated" data point both now and in the future. Downscaling can also be done in the time dimension, for example, by turning daily averaged precipitation into hourly precipitation by applying a typical daily cycle of rainfall from observations (*temporal downscaling*). Though these spatial and temporal modifications to the model output provide characteristics that appeal to some practitioners, the impact of modifications on the uncertainty description of model projections is complex. Hence, their contribution to usability of model projections and the influence on science-based credibility is controversial.

Climate model simulations may contain upwards of 100 variables that are used by climate scientists to understand model processes and their evaluation through comparison to observations. The most widely used downscaled datasets usually provide a small set of variables compared with the original model simulation. Most often temperature (mean, daily maximum, and daily minimum) and mean precipitation (daily and monthly) are provided. However, the variables (and time frequency) most salient to practitioners' applications are, often, not immediately available.

Many practitioners are looking for derived values that have well-established sensitivities to weather, usually called an **indicator** or **index**. For example, many applications are sensitive to cumulative measures of warm and cold temperature (heating and cooling degree days, or heat index) and precipitation (frequency and intensity), or lack thereof. Other applications are sensitive to a particular temperature threshold, correlated with, say, a particular time in the growth cycle of a plant or animal. Indices, which often measure persistence and variability, bring attention to modes of variability, for example the Arctic Oscillation or El Niño, and how those modes will change. The diversity of these indices is enormous, and it should be presumed the users of climate projections would need to calculate and evaluate salient indices on their own. Another alternative is to work with interpreters or climate model experts themselves. Just as daily maximum and minimum temperatures can be produced from models, indices can be produced from models. One example might **heating and cooling degree days**. A heating degree day is a measure of each day when the daily average temperature deviates (colder for heating, higher for cooling) from a standard (usually around 65 °F or 18 °C), and represents the cumulative energy demand for keeping buildings in a "comfortable" range. Heating or cooling degree days can be produced while the model is running and saved. But this requires early discussion between those running the model and those using the model, which is hard to achieve.

12.2.3 Climate Assessments

A formal interpretive instrument meant to enhance the usability of climate knowledge, including model projections, is the assessment. The best-known climate assessments are the ones from the Intergovernmental Panel on Climate Change (IPCC).[4] These assessments address the physical science basis of climate change (discussed in Chap. 11), evidence of impacts, and the state of mitigation of emissions and adaptation to present and future climate change. The IPCC makes great effort to define and codify the discussion of uncertainty. There is formal communication in three different working group assessments of (1) climate change science, (2) climate change impacts, and (3) responses for policy makers. Climate models are generally run and evaluated for the first working group (science) and used to estimate (2) impacts and (3) possible policy responses.

Many countries perform assessments themselves to refine usability for a particular region and, increasingly, provide services to improve the usability of climate projections and knowledge. The 2014 National Climate Assessment for the United States[5] puts substantial effort in the development of online information to address some of the issues of provenance and usability discussed earlier. Most other developed countries conduct similar assessments, relying on national climate scientists and policy analysts to interpret the global knowledge and assessment regionally. These assessments rely on the same climate model output, often supplemented with regional climate model output to provide detailed assessments for even parts of countries. Therefore, there is progress in the development of a chain of

---

[4]Intergovernmental Panel on Climate Change, http://www.ipcc.ch.
[5]National Climate Assessment, http://nca2014.globalchange.gov.

information, and this is a good source of translation of climate model results into more salient form.

12.2.4   Expert Analysis

Finally, many practitioners look to expert analysis when faced with the complexity (logistical and science-based) and barriers to using climate projections and observations. Practitioners often desire and use fact sheets, narrative and graphical summaries, narrative judgment, guidance, and advice. These products are anchored in climate observations and projections, and the organizations that produce them are effectively "translators" who have substantial expertise in the language and practice of climate science.

An important part of the interpretation of climate projections is the discussion and description of uncertainty. In many successful examples of climate-change planning and management, uncertainty quantification is not necessary. But a salient and qualitative understanding of uncertainty is almost always necessary. Chaps. 9 and 10 have discussed uncertainty extensively in the context of evaluation of climate models for particular uses, and Chap. 11 discussed evaluating confidence in projections. The iterative exchange of information and knowledge among data/knowledge providers and users places the climate uncertainty in context with other sources of uncertainty. Therefore, the scale of the uncertainty, along with an assessment of the state of the knowledge, becomes the essential distillation of climate uncertainty in problem solving.

12.3   Uncertainty

Virtually all discussions of the use of climate data introduce the word *uncertainty* early in the narrative (in this book, see the fourth sentence of Chap. 1). Uncertainty is perhaps the dominant focus. Uncertainty is present in many forms. As we have discussed, uncertainty can be related directly to initial observations (initial condition uncertainty), to the physical climate model (structural uncertainty), and to future emissions (scenario uncertainty). There is also uncertainty from observations used in evaluation (usually a part of structural or model uncertainty).

There is another class of uncertainty about the impacts of the resulting climate projections on the anthroposphere. Impacts of climate change on human systems include impacts on agriculture, built infrastructure, and ecosystems. These impacts are often highly dependent on the local facts and details, including policy and management practices that are in place. There is uncertainty associated with the response of people. The response is often discussed in terms of changes to technology and future energy systems to reduce emissions. This is called *mitigation*. However, on a local scale, there are decisions about land use, policy, management, and engineered systems that strongly affect vulnerability and risk to weather and climate change. This is called *adaptation*.

There is also uncertainty associated with lack of knowledge and uncertainty that comes from different interpretations of knowledge. In problem solving, uncertainty associated with ambiguous definitions of terms often emerges and becomes amplified when the same terms are used in different fields of discipline and practice. This work contains an extensive glossary in an attempt to limit this uncertainty. Many of the terms used in climate modeling and climate science, such as *positive feedback*, have different connotations and meanings in popular usage or other fields.

There is a significant body of work on uncertainty that is driven by climate scientists as well as that associated with experts who study uncertainty as a discipline. In the use of climate projections, it is essential to introduce the entire portfolio of important uncertainties early on in problem solving. By bringing uncertainty to the front of the analysis, the articulation of climate uncertainty in the context of a specific problem is often simplified. For example, if the application is snowpack, then specific uncertainties about temperature thresholds for formation of rain or snow need to be assessed. We will not attempt a comprehensive review of uncertainty, because the dimensions are specific to many problems. The fact that uncertainty changes for each problem is the critical statement. Instead, we focus on the uncertainty usually associated with climate model projections.

Chapter 10 divided uncertainty into initial condition (internal variability) uncertainty, structural (model) uncertainty, and scenario uncertainty. On the timescale of a century, scenario uncertainty dominates ($\sim$80 %) on a global scale.[6] Model uncertainty is $\sim$20 %, and internal variability is very small. With the reduction of the spatial scale of interest, on the century timescale, the internal variability is still small ($\sim$10 %), but model and scenario uncertainties are comparable in scale. This scale dependence of uncertainty, in which the uncertainty becomes more difficult to define at smaller spatial scales, is a recurring characteristic and is important to establishing salience for use of climate projections.

On shorter timescales, often relevant to planning 20–50 years into the future, the three sources of uncertainty are more equally partitioned between model response and initial conditions. On this timescale, there is not much difference in the model response to different emissions scenarios (the representative concentration pathways, or RCPs, used for climate model projections). Therefore, in many applications, emissions scenarios in the near term do not provide significantly different climates. For spatial scales smaller than global, internal variability dominates uncertainty in the first decade or two. Following those first decades, the model uncertainty and the internal variability are comparable, with scenario uncertainty ultimately assuming a major or dominant position after 50 years.

This partitioning offers information for the practitioner. Notably, the relative importance of uncertainty at different timescales is revealed, which potentially

---

[6]Hawkins, E., & Sutton, R. (2009). "The Potential to Narrow Uncertainty in Regional Climate Predictions." *Bulletin of the American Meteorological Society, 90*(8): 1095–1107.

simplifies the range of choices important for a particular problem. Many planning and management activities have small spatial scales and timescales of the next three to five decades; thus, the choice of emissions scenario is less important and the representation of internal variability is more important. Large model uncertainty argues strongly for intensive specific evaluation of model uncertainty (e.g., if the issue is snowpack, is the current temperature correct in the model?) or even the use of multiple models (multiple-model ensembles).

Scenario uncertainty is also associated with the impacts side of the problem. Recall the example of the impact of tropical cyclones on a coastline. That impact will change over time with the built environment, as well as with changes in cyclones. So the scenario for practitioners may also include aspects of the human or natural system impacts that are not treated by the climate model. One difficulty is consistency: If impact scenarios are estimated, like the built environment on a coastline, they should be consistent with the climate scenarios. The human side of the problem may matter between different scenarios, even if the climate projection is similar, as in the case for the example of the built environment around a coastline. Even if cyclones do not significantly change in 30–50 years, the built environment might very well change. The human narrative in scenarios for climate prediction is evolving in 2015, and the current versions of climate scenarios (beyond the RCPs that predict emissions) are called **Shared Socioeconomic Pathways** (SSPs).[7] These scenarios contain not just emissions, but also the growth assumptions used to estimate the emissions, and a narrative describing the assumptions about the future of society.

Initial condition uncertainty is most important when the goal of the simulation is to represent actual "climate forecasts" rather than representative "climate projections." Projections are often conditional: Given a set of emissions, the expected result is a specific climate. But a forecast is more specific. What will happen to the climate in 2020, or 2055 (the latter is more of a projection). The deterministic weather forecast problem is a classic example of a problem that relies strongly on accurate and complete initial conditions. In climate applications, the early decades of a simulation depend strongly of the initialization of the ocean. It is possible for the same model to determine quite different climates from different initializations, which is a motivation for ensemble results. Even if the initializations were (impossibly) perfect, model errors would lead to imperfect forecasts.

There are methods for mathematical quantification of uncertainty. These methods for uncertainty quantification involve understanding how perturbations to the different sources of uncertainty change the results. The computational demands of climate models as well as the complexity make brute-force methods of uncertainty quantification impractical. The uncertainty in a few parameters can be assessed explicitly, by varying different parameters over a range. This is hard to do with

---

[7]O'Neill, B. C., Kriegler, E., Riahi, K., Ebi, K. L., Hallegatte, S., Carter, T. R., et al. (2014). "A New Scenario Framework for Climate Change Research: The Concept of Shared Socioeconomic Pathways." *Climatic Change, 122*(3): 387–400.

multiple parameters since all combinations must be tested. However, this is only one dimension of uncertainty (part of **parametric uncertainty**).[8] This point is amplified by the fact that there is no unique way to parameterize a process (see Chap. 4 for discussion); that is, the expert judgments of the model builders differ from model to model, indeed, from model configuration to model configuration.[9] This is another motivation to use ensembles of model projections, which brings attention to the statistical attributes of model performance as a primary measure of uncertainty.

12.3.1 Ensembles

There are three typical types of climate-model ensemble projections. One is an ensemble of different models with the same configuration, each running the same scenario. This is designed to focus on structural uncertainty in the models. Initial condition uncertainty is present, but it goes away for long experiments (over 50 years). It explicitly removes scenario uncertainty. The second type is a set of ensemble simulations with the same model and the same scenario that start with slightly different initial conditions to sample the initial condition or internal variability uncertainty. This explores the possible states in a single model configuration, eliminating structural and scenario uncertainty. The third type of ensemble focuses on scenario uncertainty, for example, by running the same model for more than 50 years to remove model and initial condition uncertainty. All three types are used in climate analysis. Which type is used depends specifically on the application. For example, scenario uncertainty need not be treated on 20- to 50-year time horizons but dominates in the longer term. Using these different techniques leads to the conclusion that on the century scale scenario uncertainty is the largest uncertainty, not model uncertainty.

12.3.2 Uncertainty in Assessment Reports

A leading effort to describe uncertainty in a way that is potentially usable by practitioners is associated with the IPCC assessment reports. Since the year 2000, the IPCC has provided guidance to the writing teams to develop a controlled

---

[8]Tebaldi, C., & Knutti, R. (2007). "The Use of the Multi-Model Ensemble in Probabilistic Climate Projections." *Philosophical Transactions of the Royal Society A: Mathematical, Physical and Engineering Sciences, 365*(1857): 2053–2075.

[9]Schmidt, Gavin A., & Sherwood, S. (2014). "A Practical Philosophy of Complex Climate Modelling." *European Journal for Philosophy of Science* (December 9). doi:10.1007/s13194-014-0102-9.

Table 12.1 Terms and
Likelihood Estimates

Term[a]	Likelihood of the outcome
Virtually certain	99–100 % probability
Very likely	90–100 % probability
Likely	66–100 % probability
About as likely as not	33–66 % probability
Unlikely	0–33 % probability
Very unlikely	0–10 % probability
Exceptionally unlikely	0–1 % probability

[a]Additional terms (extremely likely: 95–100 % probability; more likely than not: >50–100 % probability; and extremely unlikely: 0–5 % probability) may also be used when appropriate

vocabulary and to link that vocabulary to quantitative statistical language.[10] One goal was to provide precision to terms such as *almost certain, unlikely,* and *doubtful.* Table 12.1 is from the IPCC supporting material for the 5th assessment report,[11] and it duplicates Table 11.1.

The efforts by IPCC to communicate uncertainty help to define the credibility and legitimacy of the entire body of scientific knowledge.[12] With regard to salience, the IPCC reports are most relevant at global scales and after several decades of greenhouse gas warming: They focus on scenario uncertainty and model uncertainty. These are the more certain projections from models discussed in Chap. 11. The salience or relevance of these reports is frankly inadequate for the needs of practitioners working at spatial scales on the size of watersheds and cities, and/or with planning times of 10–50 years. Further, more detailed analysis working with "interpreters" is necessary in these cases.

12.4   Framing Uncertainty

In practice, uncertainty takes on many different roles in the deliberations of teams tackling climate change problems. In many problems, the first role of uncertainty might be to reinforce political, financial, or belief positions of stakeholders, perhaps serving as a barrier to inclusion of climate change knowledge in the

---

[10]Moss, R., & Schneider, S. H. (2000). "Uncertainties—Guidance Papers on the Cross Cutting Issues of the Third Assessment Report of the IPCC." *World Meteorological Organisation*: 33–51.

[11]Mastrandrea, M. D., Field, C. B., Stocker, T. F., Edenhofer, O., Ebi, K. L., Frame, D., et al. (2010). *Guidance Note for Lead Authors of the IPCC Fifth Assessment Report on Consistent Treatment of Uncertainties.* Intergovernmental Panel on Climate Change (IPCC), 2010. Retrieved from http://www.ipcc.ch/pdf/supporting-material/uncertainty-guidance-note.pdf

[12]See Yohe, G., & Oppenheimer, M. (2011). "Evaluation, Characterization, and Communication of Uncertainty by the Intergovernmental Panel on Climate Change—An Introductory Essay." *Climatic Change, 108*(4): 629–639.

problem-solving environment. A common-heard refrain is, "If impacts are uncertain, then nothing should be done."

Generically, this use of uncertainty to position stakeholders needs to be understood by climate scientists and climate-science interpreters. The argument is often made that the reduction of uncertainty is needed to overcome barriers to action. However, there is little evidence that reducing uncertainty yields better policy outcomes or decisions.[13] Climate science and climate modeling is a science of increasing complexity, and reduction of uncertainty in a quantitative sense is unlikely. The reduction in uncertainty comes from adding more complexity and gaining more certainty about the answer, but not necessarily by reducing quantitative uncertainty: It is increasing confidence in the answer and confidence in the number of different types of uncertainty that can be addressed. Furthermore, given the role of uncertainty to bolster stakeholder positions, uncertainty can always be used to breed doubt. Therefore, it is a fallacy to maintain that reduction of uncertainty is the key to improving usability of climate projections.

Fortunately, the successful use of climate projections in planning often does not require the strict quantification of uncertainty. Practically, complex specifications of uncertainty add another level of expertise that must be interpreted, and the incremental changes to already highly uncertain parameters are not of sufficient value to justify the cost of the additional expertise. Complex specifications may make it harder to interpret climate projections by a broad community of users.[14]

Though some practitioners desire quantitative measures of uncertainty, for many people "uncertainty narratives" are all that is required to justify incorporation of climate change into planning and management. Uncertainty narratives can be framed in different ways for different problems. One productive way is to frame the uncertainty in the context of known vulnerabilities to weather. If there is an already-observed climate trend of important weather events (for example, extreme precipitation), and if that trend is consistent with model projections, then uncertainty can be discussed in relation to known weather vulnerabilities. This brings attention to the climate model's ability to represent weather features. For example, if a climate model does not represent the spatial and temporal organization of severe thunderstorms generating large amounts of rain in the central United States during summer, then it is difficult to substantiate uncertainty descriptions in regard to changes in this phenomenon. In the case of severe thunderstorms, most climate models are missing a key process (hail formation) at a subgrid scale they cannot represent.

It is also true, in many applications, that the availability of water is dominated more by policy and built infrastructure than by precipitation: "water flows uphill towards money"[15]. The spatial scale important to the water supply of a megacity is

---

[13]Lemos, M. C., & Rood, R. B. (2010). "Climate Projections and Their Impact on Policy and Practice." *Wiley Interdisciplinary Reviews: Climate Change, 1*(5): 670–682.

[14]Tang, S., & Dessai, S. (2012). "Usable Science? The UK Climate Projections 2009 and Decision Support for Adaptation Planning." *Weather, Climate, and Society, 4*(4): 300–313.

[15]Reisner, M. (1993). *Cadillac Desert: The American West and Its Disappearing Water.* New York: Penguin.

likely to include watersheds of much greater spatial scale than the city and likely to be very far removed from the city. Climate uncertainty is one piece of information input to the policy process to help determine the built infrastructure. Other inputs would include population changes and the economics of building and maintaining the infrastructure: reservoirs, aqueducts, and pumping stations. The projections from climate models might be of sufficient certainty to motivate policy changes, such as managing seasonal runoff from high mountains in order to benefit human and natural systems. The long lead times to form, approve, permit, and implement a water system allow for both the accumulation of additional observational evidence as well as the improvement of models—to inform actual specification of evolving infrastructure.

Each problem has its own unique requirements on uncertainty, and these requirements can simplify the inherent complexity of the uncertainty sources. At least the analysis can reveal the key uncertainties. For example, a problem to be addressed in the next two decades, with a solution needed to function for the two decades after implementation (i.e., a lifetime in the next 40 years), has relatively little sensitivity to the carbon dioxide emission scenario. A problem requiring specific knowledge of Arctic sea ice in the next 20 years relies on model components that have strong sensitivity to the initial state (ocean currents), a rapidly changing physical environment (melting ice changes radiative forcing), and complex multi-scale physics that are not especially well represented (see discussion in Chap. 11, and Fig. 11.6).

The more specific the application, the easier it might be to characterize the key uncertainties. If the application is to estimate sea ice to determine the feasibility of shipping routes in certain seasons in the Arctic, then the key features are narrowed to a season, and perhaps a particular threshold (sea-ice thickness less than some threshold for which an icebreaker is available). This might lead to specific uncertain processes that govern sea ice in particular seasons. Instead of looking at all available model simulations with large uncertainty (see Fig. 11.6), a subset of models could be used. The subset of models would contain those models with a good current sea-ice thickness in a particular season.

Therefore, a productive way is to step back and focus on the state of the knowledge in Fig. 11.1. This ultimately relies on fundamentals of the scientific method—observations, theory, simulation; the emergence of consistency among these three pillars; and reproducibility by many investigators coming from different approaches and scientific techniques. The state of knowledge is different for different processes and phenomena.

It is well established, for example, that sea level will rise as the planet warms up. Our knowledge of the amount of sea-level rise remains incomplete (see Chap. 11). One way to narrow the range of knowledge is to consider how fast ice might melt and increase sea level. Estimates of the physical timescale can be combined with timescales of planning, building, operations, and maintenance to narrow the range of the incomplete knowledge. In this scenario, competing explanations are of little consequence to practitioners. Placing the problem in context allows evaluation of the sources of uncertainty and whether or not there is adequate information for defensible decision making.

12.5 Summary

The key to using climate model output is understanding the credibility of the model, the legitimacy of the model, and the salience or relevance of the projection. Credibility comes from the type of model and the model development process. Legitimacy comes from a detailed understanding of uncertainty in all its dimensions for a particular problem. Salience (relevance) comes from understanding of a particular problem. Climate model output must be made relevant, and having translators or interpreters familiar with a particular application has been effective in many cases.

Climate information in many cases is just one dimension of a problem, and it may not be the dominant dimension of uncertainty, particularly where the human sphere is involved. It may matter more how society changes than how climate changes to determine the load on a particular resource (e.g., water, land). So climate model output must be put into perspective, and uncertainty assessed against particular problems to determine salience.

The use of interpreters and a focus on salience allows the dimensions of uncertainty to be reduced. These dimensions are different for particular problems. The prediction problem determines the timescale, and that determines the balance of scenario, initial condition, and model uncertainty. It may also help determine how to construct an ensemble of models for a particular problem. And the particular impacts determine what portions of model uncertainty are most important. Examples of particular aspects of model performance include intense summer convective precipitation over a region, or a particular mode of climate variability like tropical cyclones, the Asian monsoon, or blocking events. Focusing on a particular process allows a better assessment of uncertainty in a particular model, or an ensemble of models. This can be done with specific observations. It also helps to fit the model output into a particular problem, and getting the particular data on the right spatial and time scales.

We hope that this approach is useful in helping to frame the problem of assessment and use of climate models from broad uncertainties to specific and more tractable uncertainties. These uncertainties can be qualitative or, when narrowed sufficiently, even made quantitative.

For most problems, if framed in this way, it is not necessary to wait to use a future model with reduced uncertainty, and the "best" model or set of models may be different for different processes. Climate models provide a wealth of salient information that is ready to be interpreted and assessed to make specific projections.

Key Points

- Climate uncertainty may be a small part of decision making.
- Perfect models and perfect projections are not necessary for applications. Uncertain projections have value.
- Critical uncertainties may be different for each application of climate model information.
- Focusing on the particular application is one way to better understand uncertainty in climate models.

Chapter 13
Summary and Final Thoughts

In this book, we explained the basic principles behind climate models (Sect. 13.1). We described in a qualitative fashion the mechanics of how the different components of a climate model are constructed (Sect. 13.2). In the process, we focused on critical aspects of the climate system that make the different pieces complex, uncertain, and interesting. For most parts of the earth system, important mechanisms for how climate works are not necessarily intuitive. Finally, we laid out some of the methods for evaluating models, and examined what climate models are good for, and what they are not good for (Sect. 13.3). This included a detailed look at uncertainty, and a look at the applications of models for decision making.

This chapter sets out to synthesize the key points from the preceding chapters. The synthesis includes a summary of what is understood about predicting climate and what is uncertain. We also comment on future directions for climate modeling.

13.1 What Is Climate?

The goal of climate prediction is to be able to estimate and understand the present and future distribution of weather states. This distribution determines the probabilities for a weather state occurring. Climate extremes (high temperatures, periods with low precipitation) are generally low-probability events on the edges of the distribution. Climate extremes are what we really want to know about. Extremes are where the impacts are. No one is killed by the global average temperature. Fundamentally, weather and climate models are similar, but they are aiming at slightly different aspects of the system. For weather models, initial conditions are the key, whereas climate models over long time scales of a century should be independent of the initial conditions.

The climate system is a system of balances of energy and mass of air, water, and important trace compounds. The energy in the climate system ultimately comes

© The Author(s) 2016
A. Gettelman and R.B. Rood, *Demystifying Climate Models*,
Earth Systems Data and Models 2, DOI 10.1007/978-3-662-48959-8_13

from the sun. The earth absorbs sunlight mostly as visible light (shortwave energy), and radiates it back to the atmosphere and space as heat (infrared or longwave energy). Greenhouse gases alter the flow of energy in the atmosphere and trap some of this radiated heat. Water vapor (H_2O), carbon dioxide (CO_2), and methane (CH_4) are critical greenhouse gases. Humans affect water vapor only indirectly. Water and carbon flow through the components of the earth system, and much of the complexity of the climate system comes from the fact that these compounds (CO_2, H_2O) also directly alter the total energy input of the earth. Interactions and transformations of compounds across the climate system lead to many cycles. These cycles evolve on many timescales from seconds to millions of years. These cycles involve feedbacks where changing one part of the system, such as temperature, affects another part of the system, such as the amount of water vapor in the air. The reaction then alters the system, since water vapor is a greenhouse gas that further changes temperature.

Understanding the coupling of the different parts of the climate system with feedbacks is critical to understanding the future evolution of the earth's climate. Feedbacks are a key feature of large climate models. By including representations of critical processes, we try to represent these feedbacks and hence project the future state of the climate system.

The climate system is changing, and it is changing due to human activity. Greenhouse gases, mainly CO_2 and CH_4, have been increasing over the past 60 years observed from direct measurements, and for the past 150 years or so from observations of air trapped in ice cores. The chemical (isotopic) composition of the CO_2 in the atmosphere tells us that the additional CO_2 comes from fossil fuels, because the atmospheric composition of carbon (the balance of carbon isotopes) is looking more like dead plant material.

Since increasing greenhouse gases trap more energy in the system, the energy has to go somewhere. By understanding and representing the energy flows in the climate system, climate models seek to figure out where the energy is going, and what the impact of that change will be on the climate, or distribution of weather.

13.2 Key Features of a Climate Model

We use models all the time to predict the future. Examples include spreadsheets that try to predict budgets of money or goods. Some of these models are numerical. Climate models are usually not statistical but contain some processes represented with observed climate statistics, and equations built from physical theory. Essentially, a climate model is a giant representation of the "budget" of mass (of water, of carbon) and of energy in the climate system. A climate model is an attempt at representing the critical budgets and flows in the climate system in a way that they obey the basic laws of physics we observe all around us.

One way of describing the philosophy of a climate model is that a global climate model bounds each and every process by physical laws, starting from the conservation of energy and mass. From this constrained set of budget equations, combined with different representations of the processes (like condensation in clouds), complex results emerge. But these results have to be compatible with the physical laws (like conservation of mass, or the equations governing fluid flow on a rotating sphere). The emergent complexity is a reflection of reality.

The physical laws behind climate models are well known and observed. The most recent "new" theories are well over 100 years old. They are also the same physical laws that govern many other fields of science and engineering. The description of the motion of fluids in the atmosphere and ocean are the same equations used to build numerical models of how an airplane will perform. The equations that govern the flow of energy in the climate system from the sun, through the atmosphere to the earth, and then back are the same equations describing how cellular phones and radios work.

13.3   Components of the Climate System

Climate modeling has been enabled by the rapid increase in computer power that permits many of these relatively simple equations to be solved all together on more and more detailed grids of points on the planet. Climate and weather modeling were among of the first uses of digital computers in the 20th century.[1] More computer power has led to increases in complexity and increases in resolution (more points, smaller scale for each one). This evolution will continue into the future (see Sect. 13.5).

In constructing a climate model, a series of individual components, each representing one sphere of the system (atmosphere, ocean, cryosphere) is typically developed. Climate models started with just an atmosphere model and have grown to include oceans, land, and sea ice. Climate models now also typically include chemistry and representations of the flow of nutrients like carbon in the climate system. The flows of energy and mass, particularly of water mass, are critical for understanding climate. Climate models are models of the earth system that solve a set of dynamical equations. But there are also statistical (or **empirical**) models of climate and individual processes in the climate system. Statistical models represent climate-system processes with relationships among variables based on past observations. Representations (or parameterizations) of complex physical processes are often statistical models based on fits to observations. These are also called 'empirical' models. The danger of statistical models is that they are only as good as the observations of the system they seek to represent. If conditions change so that inputs are outside of the

---

[1]The earliest digital computers were used for estimating artillery firing tables and simulating the physics of the atomic bomb. See Dyson, G. (2012). *Turing's Cathedral: The Origins of the Digital Universe*. New York: Vintage.

observed range on which the model was built, or because of another factor not predicted, the statistical model may not be valid. The risk of going out of bounds of the data set used to develop a model is called extrapolation. As a result, statistical or empirical models are often limited in use to particular processes, or carefully used for relating climate variables to local conditions (statistical downscaling).

In all of the component models, there are equations for different transformations and processes (like clouds), and equations that govern the motion of air or water. A great deal of the complexity and uncertainty in climate models comes from processes at small scales that have to be represented by parameters rather than fundamental equations. These representations are often called parameterizations. The goal is to represent a process or set of processes in a particular component of a climate model. Sometimes parameterizations are tightly coupled to physical equations of the climate system. Other times, they are based on fitting a function to observations. These functional fits are empirical or statistical models described above. One needs to be careful of extrapolation. For example, if the representation of the size of ice crystals in a cloud is based on observations that range from 32 to $-4\ °F$ (0 to $-20\ °C$), then when the temperature is below the lower limit ($-4\ °F$ or $-20\ °C$), the values are "out of range."

Most of the problems and complexity of parameterization come from variations in the climate system at subgrid scales, that is, those smaller than the size of a single model grid box. In the example of ice crystal sizes, there is not one single size of ice crystals in a 62×62 mile (100×100 km) grid box: There are many sizes within clouds or a single cloud. The clouds may also not fill a particular volume of grid box. So there are interacting parameterizations (of the microphysical structure of clouds, and of the horizontal extent of clouds). Representing this variability at the grid scale is a central problem of parameterization. Higher-resolution models (smaller grid boxes) seek to get to the scale where the variability is not important: With small grid boxes the size of a football field (about 100 yards or 100 m), a single cloud can probably be assumed in the volume. Another emerging method for parameterization is to recognize that the state itself (i.e., the concentration of cloud drops in a grid box volume) is not constant, and instead of a number it can be a distribution: a probability distribution function of size of ice crystals in clouds in a particular large box, representing many clouds.

13.3.1 The Atmosphere

The atmosphere is the sphere that we live in, and it is highly changeable. There are several types of atmosphere models, from simple reduced-dimension models (a single column model or a simple zero-dimensional box or energy balance model), all the way up to general circulation models (GCMs). GCMs represent the entire atmospheric circulation with only top and bottom boundaries. The goal of global GCMs is to represent each point on a grid by a set of numbers (the state of the system at that point). This is the essence of a finite element model, where each grid point is an element. Ultimately, the description has three dimensions: two horizontal and one vertical. Some models just try to represent a single column, or a single box. A series of equations are solved for each point. These equations represent different processes of the system, like clouds in the atmosphere. Generally, the same concept is used across climate models for the different components, which are generally all finite element models. Atmosphere models must parameterize key processes. Key processes include the transformations of water into clouds and precipitation, the motion of air, and the flow of energy to and from the surface.

In addition to physical processes that are parameterizations, climate models must represent motions and the atmospheric general circulation. The atmospheric circulation can be described by the basic physics of a gas on a rotating sphere, with one extremely important complication: water. Water is a unique substance in the climate system, found naturally in the atmosphere in all three phases: water vapor gas, liquid water, and solid ice. Water is critical in most parts of the climate system. In the atmosphere, it plays a critical role in storing heat used to evaporate it, and releasing heat when it condenses.

The other critical complexity of the atmosphere is the range of scales that are important. The patterns of wet and dry regions are determined at the global scale, but important aspects of how, and when, water condenses occur on scales of a fraction of a millimeter. The range of scales in the atmosphere is a critical problem. The problem is the worst when the scale of interest is close to the grid scale of the model. When the important scale is large, then the model can represent it with one value for each grid box (like the general circulation). When the scale is small, such as a cloud drop, in a large grid box, the billions of drops can be represented statistically (as a distribution of drop sizes). But when the scale is intermediate, such as for clouds and cloud systems that may be 1–20 miles (2–32 km) in size, the scale cannot be represented well statistically. In a single grid box, there are too few clouds to use statistics to represent them, but since a number of different clouds may exist within a grid box, using a single value is not an ideal representation either. Ongoing research is currently underway to better model phenomena at intermediate scales.

13.3.2   The Ocean

The ocean has a similar hierarchy of modeling tools, from simplified versions that just provide a "wet blanket" under the atmosphere to complex models of the ocean general circulation (ocean GCMs). The ocean circulation is driven by surface winds and by buoyancy forces due to changing density (much like the buoyancy in the atmosphere that creates clouds). The density of water changes with the temperature and salt content, so both temperature and salinity can affect the circulation of the ocean. The ocean has a mixed layer that exchanges rapidly with the surface, and a strong density gradient outside of polar regions beneath this mixed layer, which separates the upper ocean from the deep ocean.

The ocean circulation is a complex result of these wind and buoyancy forces, acting on a rotating planet with ocean basin boundaries. The currents we see are a consequence of the combination of these forces. Salt content (salinity) is regulated by evaporation of water in the tropics and the formation of sea ice in the polar regions, leaving the salt behind in the ocean. Salt content is also regulated by the input of fresh water from the land surface (rivers) and directly from precipitation. Sea ice is also important for changing the reflectivity (albedo) of the surface, and insulating the ocean from a cold polar atmosphere. The ocean is a large reservoir of heat and a large store of carbon. These reservoirs play a large role in regulating the climate of the earth on long timescales. Currently, it seems that some of the heat being absorbed by the planet is going into the deep ocean and not warming the surface. That is like a "debt" that will eventually be paid in higher surface temperatures when this heat gets released. The timescales of the ocean circulation are long, and water that sinks to the deep ocean may not see the surface again for many hundreds of years.

Like the atmosphere, parameterization of key processes is important in the ocean, and often hard to represent due to subgrid variability. There is small-scale, buoyancy-driven vertical motion that is hard to represent. And a significant fraction of the oceanic heat transport occurs in small-scale eddies (loop currents) that may not be resolved by global ocean model grid spacing.

13.3.3 Terrestrial Systems

While the ocean is a huge reservoir of heat and a giant regulator of climate, the land surface is where we live, and where most of the impacts of climate are felt. The land surface, or the terrestrial system, is strongly affected by the living things on the surface (the biosphere). As with the ocean and the atmosphere, water is a critical substance for the biosphere and for regulating climate. Water fluxes are strongly affected by plants. Plants use water in respiration, bringing it up into their tissues where some evaporates in the process of photosynthesis, a process called evapotranspiration. Evapotranspiration from plants brings water from the soil up to the leaves of plants, where it can exchange with the atmosphere. This is critical for cycling moisture between the land and the atmosphere.

The growth and decay of plants also depends on critical nutrients such as nitrogen and carbon. In addition to water, carbon is the other interactive component of the terrestrial system, changing forms from solid earth to plant tissue to gas in the atmosphere.

Modeling these cycles in terrestrial systems involves representing the energy and substance (carbon, water) as it flows into and out of the system. Terrestrial models are more stationary than the atmosphere or ocean: They do not move. They describe the physical flows of the system (biogeophysics) and the plants (ecosystems) that govern and alter those flows. Ecosystems can evolve and feedback on the land surface through nutrient cycling and changes to the absorption and retention of water and heat.

Terrestrial systems also include a frozen portion: snow cover and ice sheets on land, known as the cryosphere. The cryosphere is important for altering absorption of solar radiation and changing surface fluxes. Snow cover is also an important seasonal part of the climate system for the water available to humans: Snow changes the timing of runoff by storing water on the land that can be released later. Ice sheets also store water that affects sea level. Greenland represents 23 feet (7 m) of sea-level-equivalent water, and Antarctica ~ 230 feet (70 m). That matters a lot to the 600 million people living in low-lying coastal zones.[2]

Finally, terrestrial systems also include human systems. Some physical climate models (especially simple ones) are being coupled to economic models that can simulate human systems, and so generate predictions of future climate that include the feedbacks of human societies on the climate system. One of the biggest human feedbacks is how much CO_2 we emit to the atmosphere. Another human feedback is changes that society makes to the land surface (e.g., removing forests for cropland).

13.3.4 Coupled Components

All of these components are coupled together in a comprehensive climate model. Coupling involves testing component models with observations (see below) and then attempting to put them together. The coupling layer is sort of a clearinghouse that passes information between components and reconciles their "accounts" of mass and energy. For example, a model of the ocean is usually developed by forcing with observed winds and temperatures at the ocean surface. An atmosphere model is usually developed with fixed-surface ocean temperatures. If coupling is done appropriately, then the climate should not have surface temperatures that drift over time, if energy and mass are conserved. This has taken a while to get to work properly, and one of the big advances of climate modeling in the past 20 years has been the ability to couple appropriately the atmosphere and ocean and achieve a balanced and stable global climate. The complex interactions among components of the climate system make diagnosis of coupled models difficult. But the coupling also enables evaluation of coupled phenomena across components, like the atmosphere-ocean interactions that result in phenomena like the El Niño Southern Oscillation: a pattern of changing sea surface temperatures with large-scale effect on the global distribution of precipitation. These emergent coupled behaviors are strict tests of the fidelity of models. Climate models do not parameterize phenomena like El Niño; they arise from representing basic processes (e.g., clouds, atmospheric and ocean motions) in the climate system.

---

[2]McGranahan, G., Balk,D., & Anderson, B. (2007). "The Rising Tide: Assessing the Risks of Climate Change and Human Settlements in Low Elevation Coastal Zones." *Environment and Urbanization, 19*(1): 17–37. doi:10.1177/0956247807076960.

Different types of climate models can also be coupled to each other. This is often done to use a high-resolution model in a limited area to generate high-resolution and high-frequency statistics. Variables that are outside of the limited-area model are described by a coarse model. Coupling a high-resolution model inside of a coarser (usually global) model is also called *nesting*. Nesting is often done to achieve high-resolution simulations in a particular region with limited computer resources.

13.4 Evaluation and Uncertainty

For the consumer of model output, quality is a critical question. How good is a climate model? What is a good model? Ultimately, models are fit for purpose. A good model is a model that is fit for its purpose.

13.4.1 Evaluation

So how is a good model determined? Models of all sorts are usually evaluated against some set of observations. A climate model should reproduce the present climate. Evaluation against observations is a necessary, but not sufficient condition for predicting the future. Evaluation of a model against a set of observations also requires a good knowledge of the magnitude of the uncertainty in the observations, and how comparable are the model and the observations. Evaluation also requires using the right observations and right processes to make sure the model is salient (relevant) for the intended purpose.

But reproducing observations does not guarantee a model can reproduce the future. The future response of a model may be outside of the range currently seen in the observations. This means the present is not a sufficient condition to constrain the future. A central problem of climate modeling is that we do not yet know what a sufficient condition is. We test models against observations of the recent past and present. We also continue to look for records of past climates that are preserved in various records: whether in gas bubbles from ancient atmospheres in ice cores, or in the width of tree rings over time, or in the fossilized creatures in ocean sediments. We try to expand the range of possible observations, but since the direction the climate is going now has not been seen on the planet in millions of years, inevitably we are going to have some extrapolation.

Ultimately, climate models are evaluated and compared extensively to different observations from the past: the last 100 years, all the way up to recent weather events. Climate models have a fundamental constraint on conservation of energy and mass. The global constraints, with a single boundary of the system at the top of the atmosphere model, provide powerful constraints on climate models. Few other models have these constraints (weather models usually do not). If a model conserves energy and mass, then the energy from the sun put into the system has to go somewhere. Most of the energy escapes again, but if mass is conserved, then the difference between the energy into the system and out represents energy available in the system. Figuring out where the energy goes is complex, but it is necessary to make sure energy is conserved. This also allows us to move "off scale" of current energy inputs and have some confidence that we are not accidentally gaining or losing energy in the simulated climate system.

The concept of evaluation and the energy and mass constraints can also be used to describe how a climate model is able to represent the complex earth system with complex interactions of processes occurring on many scales. If each process or parameterization or set of processes (such as a cloud model, or a biogeophysical model of how plants move water and carbon) can be evaluated against observations, and also is bounded by physical constraints, then the resulting combination of these processes should be able to represent important features of the climate system.

What does this basic physical constraint mean? For a cloud model (or cloud parameterization in an atmosphere component of a climate model), there are a series of descriptions of evaporation, formation of cloud drops, how rain begins to fall, freezing, and the like. But the overall cloud can have only as much water as is available to condense, and the energy of that condensation and/or evaporation has to go somewhere. These constraints act at every point in space and time in a model, and require all clouds in a model to meet these constraints and be physically realistic. Add up many processes pushing and pulling on the system, and climate models actually do a pretty good job of getting a decent climate for the present based on detailed comparisons to observations. The constraints of energy and mass also allow for some confidence in prediction. Another method of evaluation is to use a climate model with appropriate initial conditions to simulate individual weather events. Many models are moving to "unified" weather and climate models for this reason (see below).

13.4.2 Uncertainty

Prediction has different uncertainties over different time and spatial scales, and this distinction is critical for understanding how to use climate model output. Predicting the near term is a similar exercise to weather prediction, even if it is considered on a timescale of a season or several seasons in advance. In the short term, prediction is dominated by the uncertainty in the present state, or initial condition uncertainty. This is true on the course of a few days for weather, and maybe a few years in the atmosphere with longer-term variations in El Niño and in ocean circulation patterns. On scales of 20–50 years, the structural uncertainty in a model is important. Structural uncertainty is what we usually think of in terms of model errors. These are errors in the formulation of the model processes (parameterizations) or the interactions between processes. On spatial scales smaller than global and timescales smaller than a century, model uncertainty tends to dominate: If a model represents a process badly that is important in a particular region (like ice clouds in the Arctic), then the model is likely to have a structural bias in that region.

On longer timescales of a century, the uncertainty in human aspects of the system such as emissions of greenhouse gases dominates. The climate of 2100 is more dependent on how much we choose to emit than on the differences between different models. This is known as scenario uncertainty. That means that the climate of the end of the century is really dominated by human system uncertainty, not by uncertainty in the physical climate system. Put another way: It is our future to determine, and we do not need better climate models to make a decision on what future we want. However, to adapt to the impacts of climate change, we need to know local impacts, and local impacts are dominated by model uncertainty even at long timescales.

A common way that models are used for broad climate projections is to create a set (an ensemble) of possible realities that can be used to describe the internal variations of a model or a set of models. Ensembles can be used to provide a range of predictions or projections. A projection is dependent on things outside of the model that must be specified (such as greenhouse gas emissions). Different sets or ensembles of model simulations use different inputs, scenarios, or models. Ensembles can be used then to understand this range of uncertainties. Ensembles can be conducted with a single model. Single-model ensembles eliminate model uncertainty and explore either scenario uncertainty by performing simulations with multiple scenarios, or internal variability by focusing on a single scenario and different initial conditions. Ensembles can also be from multiple models, to focus on the model uncertainty and remove the scenario uncertainty and minimize initial condition uncertainty.

13.5   What We Know (and Do not Know)

So what do the models tell us? There are varying degrees of confidence in climate model projections. We are unlikely to be wrong on large-scale effects that are constrained by conservation of energy and mass. We are less certain of processes that do not have strict limits of energy and mass conservation. Thus, we are less certain of climate change at regional scales. If one region warms more and the next less, the average of the two may be constrained by the energy budget. But the individual regions may change a lot. Other impacts also are not constrained by conservation. One example is precipitation frequency and intensity, which are not dependent on large-scale energy and mass conservation. To produce the same amount of rain (required by conservation of water and energy) in a location, it can rain a little for a long time, or a lot for a short time. The precipitation frequency and intensity can combine in different ways to generate the same total rainfall and result in a very different climate. The least certain aspects also relate to extreme or infrequent events such as floods (local extreme precipitation), droughts (extreme periods without water), or heat waves (extreme duration of high temperatures). We are less certain about extreme events such as tropical cyclone precipitation and intensity. We are also less certain and likely to be surprised by effects with thresholds like sea-level rise to ice-sheet melting.

Practically, what does all this mean? In fits and starts, the planet should continue to warm up. Not every year or every day will be warmer than the last (because of internal variability of weather states), but over decades it will get warmer. It is hard to make the heat go away. Thus "global warming" will be nonuniform: High latitude cold regions will likely warm more because of surface albedo feedbacks resulting from melting of snow and ice cover. And there will likely be significant changes in regional patterns of precipitation. We are less certain of how this will occur, but the prediction is for very small changes in the regions of upward and downward motion, leading to more intense precipitation in the tropics, and an expansion of the semiarid regions astride the deep tropics. We also know that scenario uncertainty will start to dominate in the latter half of the 21st century, and the different path we choose for emissions (even if that path is a choice of not making a decision and doing what we are doing now) will be clear. The degree of climate change is unknown mostly because of forcing uncertainty regarding how much humans choose to emit.

Models can also be used in a more focused way to attempt to understand the smaller-scale local effects, and to provide representations of what might occur, given the above-mentioned uncertainties. The conditional forecast is a projection, rather than a prediction. Given a scenario (the condition), climate models can provide a projection. The usability of a model for a particular problem or particular impact estimate depends on whether the forecaster is "legitimate," or trusted, whether the model yields credible results compared to observations for a particular problem, and whether the results are salient, or relevant, for the problem. The latter implies "fit for purpose": The global average temperature is not a good estimate of whether a model is fit for a particular application. The ability of a model, for example, to reproduce tropical cyclones is likely a better measure of salience for projecting possible changes in tropical cyclones (but not for Arctic climate).

Climate models are just one piece of information for decision making. Climate models are one input for a knowledge system, such as a precipitation or stream flow record for a water management system that has to simulate water storage and runoff, with both physical assets (like rivers, canals, dams, and drainage basins) and human requirements for water storage and water flow. In practical terms, climate model projections are a small piece of a complicated puzzle. When they are a very different or uncertain piece, the models become difficult to use. Understanding the uncertainty in model results is critical for making them usable and relevant. Focusing on a particular result and the processes that drive the result is one way to reduce the many dimensions of uncertainty.

Planners and decision makers need interpreters or translators for climate models who can assist them in understanding the usability of particular types of model for a particular problem. Think of it as shaping the model output to fit as a piece of the overall puzzle. One goal of this book is to engage the reader to learn more about climate models, enough to be an interpreter for a set of disciplines to help shape the interpretation of model output.

13.6   The Future of Climate Modeling

We have discussed what climate models are, and what climate models can do and cannot do. Where is the development of climate models headed? Climate-model development is an iterative process. Models respond to scientific questions and needs of users, or in the absence of proper interpreters, to perceived needs of users.

The current generation of climate models typically has a series of components coupled together in various ways for various scientific tasks. The core models of the atmosphere and ocean are run at different resolutions, and with different additional components as different science questions are needed. For example, detailed models of chemistry may be run to understand air quality near the surface, or to study the evolution of the stratospheric (upper-atmosphere) ozone layer. Greenhouse gases like CO_2 are often specified by concentration over time in scenarios. But detailed carbon cycle models can be used to simulate future emissions and flows of carbon and predict, instead of specify, greenhouse gas concentrations. It is rare that all model components are turned on at once, and not every model has all the pieces. This means that particular models and particular configurations of models are most relevant for different problems.

So where are climate models headed? Increased computational power drives the ability to do more computations. There is an ongoing tension between using these computations to have higher resolution and smaller grid spacing, or adding processes and components to the model to represent more processes or improve the representation of existing processes. Process improvement means representing individual climate processes (clouds) better, and this requires improved understanding and improved observations. This also applies to additional processes that need to be represented in models.

Over time, models have grown in complexity as new processes are understood, and as computational power has increased. Adding complexity and resolution requires more computational power. And because models are multidimensional, performing calculations in three spatial dimensions and the time dimension (four dimensions, total), increasing resolution by a factor of 2 means a factor of 2×2 in the horizontal, and often a factor of 2 in time. Vertical resolution may also change, adding another multiplier. So doubling resolution often requires a factor of 8 or more in computer power just because of the increasing number of grid points in all directions, and the need to take smaller steps forward in time.

13.6.1   Increasing Resolution

Models are typically run at different scales: Finer-scale models, sometimes regional climate models, are used to try to represent extremes better with fine resolution. Global models have the benefit of a self-consistent energy balance. Currently, models are typically run for century timescales at about 62 miles (100 km) horizontal grid spacing. Shorter runs for climate (many years, occasionally a century) can be run at 15-mile (25-km) scales. In a few years from 2015, the 15-mile spacing will be more typical. Current model experiments are being run for short periods or for forecasts at ranges as small as 1–8 miles (3–12 km). These experiments are often short (or just weather forecast experiments of a few days) and experimental for now. This is the range of scales at which weather forecast models are typically run.

Why the drive to increase resolution? One goal is to reduce the variations within a grid box. As the scale gets smaller, there are fewer sources of variability. One known source is terrain. Higher resolution models can better represent complex terrain and even the subtle effects of gentle terrain (which may preferentially organize thunderstorms, for example). Another goal of higher resolution modeling is to reduce the number of processes that need to be parameterized because their scale is smaller than the grid spacing, and to represent those processes more explicitly. A smaller grid box of 1–8 miles may not need to be "partly cloudy"; perhaps it can be all cloudy, and the adjacent box clear, while a larger region representing both boxes would be "partly cloudy."

Some processes will remain parameterized (like the distribution of cloud drops whose size is the width of a human hair), but it is hoped that many of these processes are well separated from the grid scale and can still be treated statistically. Other processes, like the dynamic updrafts in clouds, or the organization of such updrafts into large storm systems, have scales from 1 to 8 miles. As models get to higher resolutions, these processes approach the grid scale, where they may not be well represented explicitly but they are hard to parameterize. This has become known as the "gray zone," because how to treat many important processes is not clear. There are many gray zones in climate modeling, but perhaps the one most people refer to is the regime between 1 and 8 miles (3–12 km), which corresponds to a complex cloud scale.

Higher spatial resolution enables unification of regional climate models and global climate models: Regional scales can be simulated with high-resolution global models. These can be either uniform-resolution or variable-resolution grids that focus on a particular region. These variable-resolution grids can be nesting two separate models, where one is on the large-scale grid and is used to force boundaries of a finer-scale model. Or the variable resolution can be a single uniform grid that changes its horizontal extent in different regions.

Improvement in climate models is driven by computational power. Faster computers enable more computations, with either more detailed processes or finer resolutions.

13.6.2 New and Improved Processes

Some newly developed parameterizations are evolving rapidly. Other processes have been represented in models for 30 years or more, and methods are fairly well explored. But new methods are developed all the time either for the "bulk" representation of a process in a grid cell or with a "variance" approach that seeks to represent the subgrid variability found in models. One of the simplest examples is "partial cloudiness" or cloud fraction, whereby a grid box can be "partly" cloudy, and values are kept for a clear and cloudy part. There can be multiple such sub-columns within the column of a grid box, and this can be used to explicitly represent the variability at small scales.

Models are adding new processes as they are identified and described with theory and observations. Starting from just an atmosphere, then adding an ocean and more processes, then a land surface, then sea ice, there is a constant evolution and expansion of the scope of climate models as new questions can be asked. One recent advance into a new area is the inclusion of models of land-based ice sheets coupled into climate models. This is driven by a desire to understand the rapid rates of recent ice melt. "Disturbance" models (such as the occurrence of wildfires) are being added to terrestrial systems. And there is a desire to use computational power to add complexity to representations of clouds, or chemistry in the atmosphere, and the chemistry of carbon throughout the earth system.

Another aspect of additional complexity in climate prediction is coupling with the human system. The treatment in this book is focused deliberately on the physical (and biological) climate system. Typically, humans have been seen as a forcing agent. But the scenarios to run the models need to reflect the possibilities of the human system. This is what actually moves the predictions more into forecasts. We cannot really forecast the future evolution of climate unless we can estimate the human emissions into the atmosphere. That requires predicting the future energy and transport system. To do so basically requires predicting the future human economic system. One approach to reducing scenario uncertainty is to build the carbon cycle into a model and also to build human systems into a model for a more self-consistent treatment of the atmospheric CO_2 concentrations and resulting forcing for climate models.

13.6.3 Challenges

In all of these configurations of climate models, there are challenges. The challenge for representing motions in the atmosphere is a consistent treatment as the scale varies. This is even harder for representing processes like clouds. Often, as the resolution gets finer and the grid size decreases, different approaches to representing processes are used. This usually occurs when the process in question has a scale not far from the grid scale: like large cloud systems or thunderstorms. In many cases, climate models rely on methods used for smaller-scale weather models to improve their process representations (parameterizations).

One ongoing trend is to make unified models for weather and climate prediction, using the same parameterizations and processes, but running the model in different ways for weather or climate. For weather, a system is used to initialize the model carefully with current observations, and the model is run forward for a few days.

For climate, the initialization does not matter, and the model is run for a long time. There are benefits of unified models, both to weather forecasting and climate prediction. Climate prediction benefits from the constant verification and testing against weather events (including extremes) in weather forecast models. Weather models benefit, too. They are forced to make improvements in conservation of energy and mass to run in climate mode. As weather models are starting to be run for seasonal prediction over months rather than days, conserving energy and mass and having a proper energy budget is critical.

One final note is that improvements must balance where to put increased computer power. Should a model be run with more advanced processes or finer resolution? It depends on decisions made in the development of a model, and what the aims of a model are, and the deficiencies. Different models will make different choices. When selecting climate model projections for applications, some care should be taken to select those climate models that perform well on evaluation of specific processes that are relevant to the application. Some applications benefit from high spatial resolution, and some do not.

13.7   Final Thoughts

Climate models are representations of the complex climate system. They are themselves complex constructions of the interactions of many individual processes. A typical climate model now contains as many lines of computer code as a computer operating system. The processes in climate models are governed by basic physical laws. These laws are applied at the process level, the sum of processes (component level), and the coupling between components in the climate system. The result is an emergent complexity from the interaction of these bounded processes and then the interactions between the different spheres of the climate system. Climate models attempt to represent a complete and consistent earth system and thus benefit from fundamental constraints on energy and mass. This last benefit is often unique to climate models. Climate models are therefore complex, but they are built from basic physical laws, and they do a remarkable job of simulating many aspects of the earth's climate. One of the continuing challenges is representing the many different scales of variations in the climate system that are too small to represent with a single number in a large grid box.

Sometimes, climate modeling is derided as an *art*. The term is derogatory, intended as the opposite of *science*. The implication is that climate models are a hopeless tangle of competing equations that make no sense in the whole, and that cannot hope to represent the key processes that will determine the magnitude of climate change. In particular, since the models contain numerous uncertain parameters, it is argued that there is a "hidden art" to adjusting these parameters in any model and that the process of adjusting these parameters (often called *tuning*) does not follow the scientific method.[3] But this is not really true. The laws of physics and fundamental constraints of conservation bound each process. As models get more complex, parameterizations represent processes more explicitly and are described in ways closer to physical laws, using parameters that can be constrained by observations.

The adjustment or tuning process of a set of parameters to match a set of observations is an optimization problem that can also be completed objectively. Recent attempts at quantifying uncertainty in climate model adjustment have shown that an objective algorithm reproduces the intuition of model developers.[4] This evaluation is important for putting climate models on a sound scientific footing. There is also proof of the utility of climate models from past climate model predictions. Predictions from climate models nearly 30 years ago follow well the trajectory the climate has taken,[5] much better than any economic model has done with the global economy over the past 30 years.

We are more certain of what will happen at longer time scales and larger spatial scales (global). This arises from the nature of the problem, and the transient effects of internal variations in the system. Much of the remaining global uncertainty focuses on clouds, since the response of clouds to climate changes (cloud feedback) affects the total net energy in the earth system. The role of the ocean is also critical. It is a huge reservoir of heat, and it controls where that heat goes and how much goes into the surface or how much heat the system "saves" for later.

The consequence of analyzing the uncertainty in climate model projections in this way is surprising. If we use the global-scale average surface temperature as the defining metric of *global warming*, then projections of global warming are uncertain mostly because we do not know the quantity of human greenhouse gas emissions in the future, not because of uncertainty in climate models. This is scenario uncertainty. The goal of climate models is to minimize model uncertainty to be able to make more confident projections about regional scales with high-resolution climate models or limited-area (regional climate) models.

Using climate models appropriately requires understanding many of these subtleties. Most of all, it requires an understanding of uncertainty and how to assess uncertainty in climate model projections (and the difference between predictions and projections) for a particular problem, recognizing that uncertainty will vary with the application. We hope in the end that the reader is now a more competent interpreter or translator when confronted with climate model output to use.

---

[3]For a good discussion of the methodology of model optimization, see Schmidt, G. A., & Sherwood, S. (2014). "A Practical Philosophy of Complex Climate Modelling." *European Journal for Philosophy of Science* (December 9). doi:10.1007/s13194-014-0102-9.

[4]Zhao, C., Liu, X., Qian, Y., Yoon, J., Hou, Z., Lin, G., et al. (2013). "A Sensitivity Study of Radiative Fluxes at the Top of Atmosphere to Cloud-Microphysics and Aerosol Parameters in the Community Atmosphere Model CAM5." *Atmospheric Chemistry and Physics, 13*(21): 10969–10987. doi:10.5194/acp-13-10969-2013.

[5]Hansen, J., Fung, I., Lacis, A., Rind, D., Lebedeff, S., Ruedy, R., et al. (1988). "Global Climate Changes as Forecast by Goddard Institute for Space Studies Three-Dimensional Model." *Journal of Geophysical Research, 93*(D8): 9341–9364. doi:10.1029/JD093iD08p09341.

Climate Modeling Text Glossary

For further reference, there are a number of online glossaries of climate terms. Much of the glossary here is based on these sources. In many cases, these glossaries trace back to the AMS Glossary.

1. Intergovernmental Panel on Climate Change (IPCC) Glossary: https://www.ipcc. ch/publications_and_data/publications_and_data_glossary.shtml

2. American Meteorological Society (AMS) Glossary: http://glossary.ametsoc.org/ wiki/Main_Page

3. Skeptical Science Glossary: https://www.skepticalscience.com/glossary.php

Glossary Terms (Chapter in which term appears in parentheses).

Aerosol particles (5) small solid or liquid particles dispersed in some gas, usually air.

Albedo (2) the ratio of the reflected radiation to incident radiation on a surface. Shortwave albedo is the fraction of solar energy (shortwave radiation) reflected from the earth back into space. Albedo is a measure of the surface reflectivity of the earth. Ice and bright surfaces have a high albedo: Most sunlight hitting the surface bounces back toward space. The ocean has a low albedo: Most sunlight hitting the surface is absorbed.

Anthropogenic (3) human (*anthropo-*) caused (generated).

Anthroposphere (2) also called the anthrosphere; the part of the environment made or modified by humans for use in human activities and human habitats.

Aquifer (7) an underground layer of water-bearing permeable rock or unconsolidated materials (gravel, sand, or silt) from which groundwater can be extracted using a water well.

Arable land (7) land capable of being plowed and used to grow crops.

© The Author(s) 2016
A. Gettelman and R.B. Rood, *Demystifying Climate Models*,
Earth Systems Data and Models 2, DOI 10.1007/978-3-662-48959-8

Atmosphere (2) the gaseous envelope gravitationally bound to a celestial body (planet, satellite, or star).

Baythemetry (6) originally referred to the ocean's depth relative to sea level, although it has come to mean submarine (underwater) topography, or the depths and shapes of underwater terrain.

Biogeochemical cycles (7) natural pathways by which essential elements of living matter are circulated.

Biogeochemistry (7) the scientific discipline that involves the study of the biological, geological, chemical, and physical processes and reactions that govern the composition of the natural environment.

Biogeophysics (7) the study of water and heat (energy) flow through the soil and plants on the earth's surface.

Blocking events (10) the obstructing, on a large scale, of the normal west-to-east progress of migratory cyclones and anticyclones. This anomalous circulation pattern (the "block") typically remains nearly stationary or moves slowly westward, and persists for a week or more. Prolonged blocking in the Northern Hemisphere occurs most frequently in the spring over the eastern North Atlantic and eastern North Pacific regions.

Bottom water (6) the water mass at the deepest part of the water column. The densest water in the column.

Boundary layer (5) a layer near the edge of the ocean or atmosphere. The ocean boundary layer is right below the surface; the atmospheric boundary layer just above.

Brine pockets (6) pockets of salt water of high concentration in sea ice resulting from the rejection of salt on freezing of sea water into sea ice.

Bucket model (7) a representation of the field capacity (water-holding capacity) of soil, where the soil can hold a fixed amount of water before it overflows (runoff).

Buoyancy (6) the property of an object that enables it to float on the surface of a liquid, or ascend through and remain freely suspended in a compressible fluid such as the atmosphere. Also the upward force exerted on a parcel of fluid (or an object within the fluid) in a gravitational field because of the density difference between the parcel (or object) and that of the surrounding fluid.

Carbon cycle (7) the cycling of carbon through the earth system.

Carbon cycle feedback (7) the interaction of the land surface with climate. The feedback usually implies that increasing CO_2 will allow plants to grow more efficiently, taking up more CO_2 and reducing the CO_2 increase.

Carbon sink (7) a process that removes carbon from a reservoir.

Cell (4) a grid cell (the smallest unit that is resolved) in a model. A cell is one vertical part of a column.

Chlorofluorocarbons (7) also CFCs, organic compounds that contain carbon, chlorine, and fluorine; human-made compounds that are inert and nontoxic with long lifetimes in the atmosphere. They slowly release chlorine, which contributes to stratospheric ozone depletion.

Circulation patterns (2) the long-term patterns of the flow or motion of a fluid (air or water) in or through a given area or volume.

Classical physical mechanics (4) also Newtonian Mechanics; the laws of motion of physical objects.

Climate (1) the average or distribution of weather events, typically represented by averages over long periods of time (a month or more).

Climate forecasting (1) an estimate of the future state of the climate focusing on the distributions of temperature and precipitation over longer periods of time.

Climate interpreters (12) people with knowledge of the utility and use of climate models who are able to provide a link between climate-model science and applications.

Climate model (4) a model used to make forecasts and simulations of climate. Typically related to numerical weather-prediction models. Typically designed to be run (integrated) for many years.

Climate regimes (5) classification of climate into different types; regions with similar regimes have similar climates.

CO_2 fertilization (7) the enhancement of the growth of plants as a result of increase in the concentration of atmospheric CO_2. Higher CO_2 enables more efficient transfer of CO_2 into plant tissues for photosynthesis with less water loss.

Column (4) a vertical stack of grid cells in a model at a single horizontal location.

Compensating errors (5) errors that are hidden due to offsetting or cancellation. A positive bias combined with a negative bias leads to compensating errors.

Condensation (2) the transition from a gas to a liquid, the opposite of evaporation. Specifically the phase change of water from water vapor to liquid water.

Conservation of mass (4) the principle (in Newtonian mechanics) that states mass cannot be created or destroyed but only transferred from one volume to another.

Constraints (4) rules or laws that constrain or limit the behavior of different processes.

Convection (6) mass motions within a fluid resulting in transport and mixing of the properties of that fluid. Motions that are predominantly vertical and driven by buoyancy forces arising from density gradients with light air (or water) beneath denser air (or water).

Coriolis force (6) or Coriolis effect; an effect where a mass moving in a rotating system experiences an apparent force (the Coriolis force) acting perpendicular to the direction of motion and to the axis of the rotation. On the earth, the effect tends to deflect moving objects to the right in the Northern Hemisphere and to the left in the Southern Hemisphere. The Coriolis force is zero at the equator (since an object moves parallel to the axis of rotation), and a maximum at the poles.

Coupled climate system model (4) a class of climate model in which at least two different subsystems of earth's climate system are allowed to interact. A coupled model would typically couple the atmosphere, ocean, and land, sometimes also atmospheric chemistry so that different parts interact with each other.

Coupled Model Inter-comparison Project (CMIP) (11) a standard experimental protocol for studying the output of coupled atmosphere-ocean general circulation models. http://cmip-pcmdi.llnl.gov/.

Coupling (3) the interaction of two or more items, things, or processes.

Credibility (12) scientifically trusted or believable.

Cryosphere (2) the places on the earth where water is in solid form, frozen into ice or snow.

Deep ocean (6) the region of the ocean below the thermocline.

Deformable solid mechanics (7) the branch of continuum mechanics that studies the behavior of solid materials, especially their motion and deformation under the action of forces, temperature changes, phase changes, and other external or internal agents.

Disruptive innovation (10) innovation that helps create a new market and value network, and eventually disrupts an existing market and value network (over a few years or decades), displacing an earlier technology.

Domain (5) a specific and limited region.

Downscaling (5) method used to obtain local-scale weather and climate information, from regional-scale atmospheric variables that are provided by GCMs. Two main forms of downscaling technique exist. One form is dynamical downscaling, where large-scale model output (from the GCM) is used to drive a regional, numerical model at higher spatial resolution to simulate local conditions in greater detail. The other form is statistical downscaling, where a statistical relationship is established from observations between large-scale variables, like atmospheric surface pressure, and a local variable, like the wind speed at a particular site. The relationship is then used subsequently on the GCM data to obtain the local variables from the GCM output.

Downwelling (11) downward motion of surface or subsurface water that removes excess mass brought into an area by convergent horizontal flow near the surface.

Drag (5) also called resistance; the frictional retarding force offered by air to the motion of bodies passing through it.

Dynamical core (5) the portion of a model that integrates the equations of motion. It usually determines the winds and the temperatures from a set of dynamical motion equations.

Dynamical downscaling (5) method by which large-scale model output (from the GCM) is used to drive a regional, numerical model at higher spatial resolution to simulate local conditions in greater detail.

Earth system models (4) a class of coupled climate model that is coupled to the biosphere on land and/or in the ocean.

Economic system models (7) models that simulate economic activity.

Ecosystem dynamics (7) the study of the response and evolution of ecosystems to disturbances.

Eddies (6) circular movements of air or water. In the ocean, a closed circulation system produced as an offshoot from an ocean current. Eddies are the common result of the turbulence of the ocean circulation. The corresponding features in the atmosphere are the wind currents around high and low pressure disturbances.

El Niño Southern Oscillation (ENSO) (8) a significant increase in sea surface temperature over the eastern and central equatorial Pacific that occurs at irregular intervals, generally ranging between 2 and 7 years. The Southern Oscillation refers to variations in the temperature of the surface of the tropical eastern Pacific Ocean, with warming known as El Niño and cooling known as La Niña, and in air surface pressure in the tropical western Pacific. The name for the warming comes from the Spanish word for Christ child because the warming usually occurs around Christmas.

Electromagnetic radiation (4) radiation consisting of electromagnetic waves, including radio waves, infrared, visible light, ultraviolet, X-rays, and gamma rays. Energy with the form of electromagnetic waves as well as the form of a stream of photons and traveling at the speed of light in a vacuum.

Emergent (4) a property which a collection or complex system has, but which the individual processes do not have.

Empirical model (13) a model based on observation, rather than theory. Sometimes called a statistical model.

Energy balance models (5) an idealized model focused on the energy balance, the balance between the net warming or cooling of a volume and all possible sources and sinks of energy. Energy is conserved, and energy balance models use this fact to help understand the different flows and exchanges of energy in the earth system.

Energy budget (7) the accounting for the energy balance, the balance between the net warming or cooling of a volume and all possible sources and sinks of energy.

Energy flows (2) the movement of energy in the earth system.

Ensembles (10) multiple model variations with slight changes. Ensembles of model simulations can be run with variations of initial conditions (often done for weather forecasting) to sample initial condition uncertainty, variations in scenarios for a single model to sample scenario uncertainty, or with different models to sample model or structural uncertainty.

Equilibrium (2) a state in which opposing forces or influences are balanced; a state of physical balance.

Evaluation (9) the process and practice of determining the quality and value of a model or forecast.

Evaporation (2) the transition between a liquid and a gas. Specifically the phase change of liquid water into water vapor.

Evaporative cooling (4) reduction in temperature resulting from the evaporation of a liquid, which removes latent heat from the surface from which evaporation takes place. This process is also the physical basis of why humans and other animals perspire.

Evapotranspiration (7) the sum of evaporation and plant transpiration of water from the earth's land and ocean surface to the atmosphere.

Extrapolate (9) or extrapolation; the process of estimating, beyond the original observation range, the value of a variable on the basis of its relationship with another variable.

Feedback (3) the modification or control of a process or system by its results or effects. A feedback alters the processes in the system by changing inputs depending on the output.

Field capacity (7) the maximum amount of water a soil can retain before runoff; the amount of soil moisture or water content held in the soil after excess water has drained away.

Finite element model (4) a model with dependent variables represented as a finite series of piecewise-developed polynomial basis functions (finite elements). The elements are often arrayed on a grid of points where each cell is an element.

Fixation (2) the process of making something firm or stable. In general terms for climate, making a gas into a more stable form, usually a solid.

Forcing (3) or external forcing; refers to a forcing agent outside the climate system causing a change in the climate system. Examples of forcing include volcanic eruptions, solar variations, and anthropogenic changes in the composition of the atmosphere.

Forecast (9) a prediction or estimate of the future.

Forecasting (1) an assessment of a future state of a system.

Fossil fuels (3) buried combustible geologic deposits of organic materials, formed from decayed plants and animals that are used for fuel; chiefly crude oil, coal, and natural gas (methane).

Gaia hypothesis (3) proposes that organisms interact with their inorganic surroundings on earth to form a self-regulating, complex system that contributes to maintaining the conditions for life on the planet. Originally put forward by James Lovelock.

General circulation model (GCM) (5) a global finite element model that integrates the equations of motion on a sphere to represent the circulation and weather patterns of the earth. GCMs are used for both climate and weather prediction.

Geostrophic balance (6) the balance that results from a fluid on a rotating sphere. A balance between the Coriolis and horizontal pressure-gradient forces.

Green ocean (7) a term sometimes ascribed to tropical rainforests like the Amazon, referring to the large evapotranspiration and precipitation that occurs over tropical rainforests, similar to an ocean of water.

Greenhouse gases (2) gases, such as water vapor (H_2O), carbon dioxide (CO_2), and methane (CH_4), that are mostly transparent to the short wavelengths of solar radiation but efficient at absorbing the longer wavelengths of the infrared radiation from the earth and atmosphere. They thus trap heat in the atmosphere.

Grid (4) the regular or irregular set of columns or points in a model.

Grid box (4) the same as a grid cell. The smallest unit that is resolved in a model, one part of a column.

Gross primary productivity (7) the amount of energy fixed by photosynthesis over a defined time period.

Gulf Stream (6) one of the western boundary currents of the North Atlantic and one of the swiftest ocean currents with one of the largest transports.

Gyre (6) a rotating ring-like system of mean or steady large ocean currents.

Hadley circulation (5) named after George Hadley, the Hadley circulation is a tropical atmospheric circulation that in the zonal (longitude) average features rising motion near the equator, poleward flow at high altitude above the surface, descending motion in the subtropics, and equatorward flow near the surface.

Halocline (6) a vertical salinity (salt) gradient in some layer of a body of water that is appreciably greater than the gradients above and below it; also a layer in which such a gradient occurs.

Heating or cooling degree day (12) a measure of each day that the daily average temperature deviates (colder for heating, higher for cooling) from a standard (usually around 65 °F or 18 °C); represents the cumulative energy demand for keeping buildings in a "comfortable" range.

Hindcasts (9) or hindcasting; analogous to forecasting, hindcasting is a way of testing a mathematical model. Forecasting the past. Observed inputs for past events are entered into a model to see how well the output matches the known results.

Human disturbances (7) perturbations or changes to an ecosystem that occur due to human activity, such as deforestation.

Hydrologic cycle (2) also called the water cycle, the hydrologic cycle describes the movement of water on (land), and above (atmosphere) and below (soil and ocean) the surface of the earth.

Hydrology (7) study of the movement, distribution, and quality of water on earth.

Ice core record (3) the record of a quantity (such as dust amount, or carbon dioxide trapped in air bubbles) found in an ice core; a core sample that is typically removed from an ice sheet or glacier.

Indicator or index (12) derived quantities that have a relationship to weather, such as the heat index (cumulative precipitation). Other climate indicators or indices include the state of El Niño expressed as a temperature anomaly in the Pacific Ocean.

Initial condition uncertainty (1) the uncertainty in a projection, prediction or forecast due to uncertainties in the initial input conditions of the state of the system.

Initialization (5) the process of starting up a model with a set of initial conditions before the model has been run.

Insolation (3) contraction from "incoming solar radiation". In general, solar radiation received at the earth's surface. Formally, isolation is the amount of direct solar radiation upon a unit horizontal surface.

Integrated assessment models (7) models of the earth and human system that generally include both physical and social science models that consider demographic, political, and economic variables that affect emissions of greenhouse gases as well as the physical climate system. Usually the physical system is simplified.

Intergovernmental Panel on Climate Change (IPCC) (11) the international scientific body that conducts assessments of climate change science, impacts, and policy. http://www.ipcc.ch/.

Intermediate complexity models (5) simplified models of the climate system (also know as earth system models of intermediate complexity). These models usually represent the climate by an energy balance over large regions (like an ocean basin or an entire continent) that are tied together. They are less complicated than full earth system models, but they do try to represent or specify feedbacks, so they are more complex than simple idealized models like energy balance models.

Isostatic rebound (8) the rise of land masses that were depressed by the huge weight of ice sheets during the last glacial period. *Isostatic* refers to the equilibrium of the earth's crust with the mantle underneath.

Isotopes (3) different forms of the same element that contain equal numbers of protons and electrons but different numbers of neutrons, and, hence, that differ in relative atomic mass but not in chemical properties.

Kinetic energy (4) the energy that a body possesses as a consequence of its motion, defined as one-half the product of its mass (m) and the square of its speed (v): $\frac{1}{2} mv^2$.

Latent heat (7) energy released or absorbed by changes of phases of water. Condensation and freezing release heat, while evaporation and melting require heat input.

Leads (6) open water that forms between patches of sea ice, usually due to divergence (separation of ice).

Legitimacy (12) valid, objective, fair, or free of bias.

Limited-area models (5) models that cover only a part of the earth and have lateral boundaries, such as regional climate models. Such models must be given lateral boundary conditions.

Longwave radiation (5) energy emitted at wavelengths longer than about 4 micrometers (millionths of a meter) in the infrared part of the spectrum, usually of terrestrial origin.

Mean (3) the arithmetic average of a set of numbers. Defined as the total sum of all values divided by the number of values.

Median (3) the middle value of a set of numbers listed in numerical (or algebraic) order. If an even number of values, then halfway between the middle terms.

Meridional overturning circulation (6) a system of surface and deep currents encompassing all ocean basins. It transports large amounts of water, heat, salt, carbon, nutrients, and other substances around the earth, and connects the surface ocean and atmosphere with the deep ocean.

Middle latitudes (5) or mid-latitudes; the region in both hemispheres between about 35° and 65°, usually marked by a band of westerly (eastward-blowing) winds.

Mixed layer (6) in oceanography, a turbulent region of nearly vertically uniform density that, in the case of the surface mixed layer, is bounded above by the air-sea interface and below by the transition layer or thermocline.

Mixed-layer ocean models (6) an ocean model that assumes a shallow ocean depth representative of the mixed layer (33–165 ft, 10–50 m). The models do not contain a deep ocean or its circulation.

Mode (1) the most frequent value in a distribution (highest probability).

Model (1) a representation of a process or object, by necessity simplified in some way from the original.

Model uncertainty (1) the uncertainty in a model formulation, also known as the structural uncertainty. In a numerical model, model uncertainty results from imperfect representations of different processes and their interactions.

Monsoon (8) a seasonal reversing wind accompanied by corresponding changes in precipitation, or more generally the seasonal changes in atmospheric circulation and precipitation associated with the asymmetric heating of land and sea.

Natural disturbances (7) perturbations or changes to an ecosystem that occur naturally, such as wildfires caused by lightning.

Natural forcing (3) a forcing agent that is not changed by humans. Examples include changes in the earth's orbit that affect the solar input, or volcanic eruptions.

Negative feedback (3) a feedback that dampens (decreases) the response to a perturbation on a system.

Nitrogen cycle (7) the cycling of nitrogen through the earth system.

Numerical weather prediction (NWP) models (1) a numerical model used to predict the future state of the atmosphere. More formally, NWP models integrate the hydrodynamical equations with numerical methods subject to specified initial conditions for a particular time.

Nutrient cycling (7) transformation of important chemicals used by plants for food from one state or one part of the climate system to another.

Observational uncertainty (9) the unknown difference between an observation and its "true" value.

Ocean circulation (6) the long-term patterns of the motion of seawater in the worlds' oceans. It includes several types of circulations at the surface (wind-driven gyres) as well as the circulations of the deep ocean (meridional overturning circulation and thermohaline circulation).

Ozone depletion (5) catalytic removal of stratospheric ozone by chlorine. In polar regions ozone depletion reactions are accelerated by the presence of polar stratospheric clouds leading to formation of a large region of ozone depletion in spring (the ozone hole), mostly in the Southern Hemisphere.

Paleoclimate (9) climate of the geological past.

Parameter (1) any quantity in a problem that is not an independent variable (the output of the model). Also used to distinguish fixed quantities in a model from the dependent variables (inputs) in a model. Usually a parameter is a fixed mathematical constant or function.

Parameterization (4) a mathematical representation of a physical process in terms of simplified parameters. Empirical parameterizations are a functional fit between observed inputs and desired outputs.

Parametric uncertainty (10) variations in model results that come from uncertainty of set parameters in a model. A given range of parameter choices will cause variations that define parametric uncertainty.

Persistent (11) lasting for a long time.

Phenology (7) the science dealing with the influence of climate on the recurrence of annual phenomena of animal and plant life such as budding and bird migrations.

Photochemical smog (5) haze produced from anthropogenic pollutants that react with sunlight. Photochemical smog of nitrogen oxides and hydrocarbons are emitted mainly by vehicle engines but can also contain particulates.

Plant functional types (7) a system that groups plants according to their characteristics; describing plant function in ecosystems and their use of resources (nutrients).

Point (4) a single location in a model, represented by a single grid cell.

Positive feedback (3) a feedback that amplifies (increases) the response to a perturbation on a system.

Prediction (9) a forecast; an estimate of some future outcome or state.

Probability distribution (1) a probability distribution assigns a probability to the occurrence of each subset of all the possible outcomes of a set of data. A probability distribution function represents a probability distribution as the frequency of occurrence of any particular value of a set of data. A normalized distribution has the integral under the curve equal to 1 so that the probability of any value is represented by the vertical axis.

Process splitting (10) a method of model integration by which each process operates at the same time based on the same state in a model and then the results are combined.

Projection (9) an estimate of the future based on current trends, or based on assumptions in scenarios.

Proxy records (9) preserved physical characteristics of the past that stand in for direct measurements. For example, the isotopic ratio of different isotopes of oxygen can be used as a proxy for temperature of the formation of ice in ice cores.

Pycnocline (6) a vertical density gradient (determined by the vertical temperature and salinity gradients) in some layer of a body of water, which is appreciably greater than the gradients above and below it; also a layer in which such a gradient occurs.

Radiative forcing (3) measure of the influence a factor has in altering the balance of incoming and outgoing energy in the earth-atmosphere system and an index of the importance of the factor as a potential climate change mechanism.

Regional climate model (5) a numerical climate prediction model forced by specified lateral and ocean conditions. Boundary conditions can be from a general circulation model (GCM) or observation-based dataset. A regional climate model simulates atmospheric and land surface processes, while accounting for high-resolution topographical data, land-sea contrasts, surface characteristics, and other components of the Earth-system. The values at the boundaries (boundary conditions) of a regional climate model must be specified explicitly. Regional climate models can thus simulate climate variability with regional refinements, but are dependent on the boundary conditions.

Representative Concentration Pathways (RCPs) (10) four greenhouse gas concentration (not emissions) trajectories adopted by the IPCC for its fifth Assessment Report (AR5) in 2014. The pathways are used for climate modeling and research. They describe four possible climate futures, all of which are considered possible depending on how much greenhouse gases are emitted in the years to come. The four RCPs—RCP2.6, RCP4.5, RCP6, and RCP8.5—are named after a possible range of radiative forcing values in the year 2100 relative to pre-industrial values (+2.6, +4.5, +6.0, and +8.5 Wm^{-2}, respectively). The RCPs describe a wide range of possible changes in future anthropogenic (i.e., human) greenhouse gas emissions.

Reservoirs (7) a supply of a substance, especially a reserve or extra supply, or a region that holds a supply of a compound or chemical in the earth system. A reservoir of water is a space capable of storing water.

Resistances (7) or impedances; retarding forces on flows of water or nutrients.

Resolution (4) the length of the finest-described scale in a model. Typically the horizontal scale (two dimensions in space) of a single grid box in a model.

Respiration (2) gas exchange between solid and gas forms. In the case of plants, consuming carbon dioxide and releasing oxygen. In the case of animals (and bugs, microbes), consuming oxygen and releasing carbon dioxide.

Rheology (6) the study of the flow of matter like liquids, or solids (like sea ice), under conditions in which they respond with plastic flow rather than deforming elastically in response to an applied force.

Ridging (6) regions of thicker ice that stick up above the surface due to compression or convergence of sea ice.

Salience (12) relevant, useful, having a prominent signal.

Salinity (6) salt content of water.

Sampling uncertainty (9) uncertainty or error introduced by having observations only at limited points and sparsely sampling an observed quantity (like temperature or precipitation).

Scenario uncertainty (1) the uncertainty in a projection or prediction of the future due to uncertainties in inputs (boundary conditions) to a model over time.

Scenarios (9) multiple, possible descriptions of what might happen in the future. Often used as inputs to models.

Sensitivity (5) the degree to which a system will respond to an input of given strength. Systems with larger positive feedbacks have higher sensitivity.

Shared Socioeconomic Pathways (SSPs) (10) a framework that combines climate forcing and socioeconomic conditions. These two dimensions describe situations in which mitigation, adaptation, and residual climate damage can be evaluated. The core is a limited set of five SSP narratives that describe the main characteristics of future human development pathways including population, urbanization, and economic development. SSPs are the starting point for the identification of internally consistent assumptions for the quantification of emissions (similar to RCPs). Different modeling tools can be used to develop quantifications of these storylines, including factors like population, economic development, land use, and energy use.

Shortwave radiation (5) energy in the ultraviolet or visible and near-visible portion of the electromagnetic spectrum (0.4–1.0 millionths of a meter, in wavelength). These are the wavelengths emitted by the sun, and *shortwave* is used to distinguish from radiation emitted by terrestrial (low-temperature sources).

Signal (10) the real portion of some observed relationship, the opposite of noise.

Simplified sea-ice models (6) also called thermodynamic ice models; sea-ice models that do not account for ice motion and deformation and instead just simulate a surface energy balance of ice.

Single-column model (5) a model that has a single, vertical dimension (a column).

Sink (5) a route by which a measurable quantity may exit a system, such as by accumulation (in a reservoir) or chemical conversion. A loss process.

Smog (5) a natural fog contaminated by industrial pollutants; a mixture of smoke and fog.

Soil moisture (7) the amount of water in the soil.

Spread (11) variation or variability across a sample, or ensemble.

Standard deviation (9) a quantity used to measures the amount of variation, or variability, in a data set. The standard deviation is the square root of the variance. It has the same unit as the data set. Variance is the average of the squared difference from the mean. Larger spread from the mean will give a larger variance and larger standard deviation.

State (4) or state variables; the basic variables that define the state of the grid cell in a model in the atmosphere, ocean, etc. For the atmosphere, the state is described by pressure, temperature, wind (kinetic energy), and concentrations or mass of trace species. The state vector is a set of these variables at every point (cell) in a model.

Statistical downscaling (5) method by which a statistical relationship is established from observations between large-scale variables, like atmospheric surface pressure, and a local variable, like the wind speed at a particular site. The relationship is then subsequently applied to GCM output to obtain the local variables from the GCM output.

Statistical models of climate (8) a climate model based on regression techniques that relate a climate response or impact to a state variable based on past observations. Sometimes called empirical models.

Storm surge (6) onshore surge of seawater due primarily to winds in a storm, and secondarily to the surface pressure drop near the storm center.

Stratosphere (5) from the Greek for "layered region." The region of the atmosphere above the tropopause that is stable with temperature increasing with height. The region where the ozone layer is found.

Stress (6) a surface force, induced for example by wind.

Structural uncertainty (10) the uncertainty in a model formulation, also known as the model uncertainty. In a numerical model, structural uncertainty results from imperfect representations of different processes and their interactions.

Subgrid variability (5) variability on scales that are less than the grid scale of a model.

Succession (7) the evolution of plant types in a region from one to another.

Surface fluxes (6) the flow of energy into or out of the surface due to net radiation, sensible heat, and latent heat.

Surface ocean (6) the region of the ocean above the thermocline.

Sverdrup balance (6) a theoretical relationship between the wind stress exerted on the surface of the open ocean and the vertically integrated meridional (north-south) transport of ocean water. The Sverdrup balance is a consistency relationship for flow that is dominated by the earth's rotation. Such flow will be characterized by weak rates of spin compared to that of the earth.

Terrestrial biosphere (2) the portion of the biosphere (the locations where life is present) that is found on the land surface of the earth.

Terrestrial system (7) the interacting set of processes that occur on the land surface of the earth.

Thermal energy (4) the heat content, heat energy.

Thermocline (6) a vertical temperature gradient, in some layer of a body of water, that is appreciably greater than the gradients above and below it; also a layer in which such a gradient occurs.

Thermodynamics (4) a branch of physics concerned with heat and temperature and their relation to energy and work. It is used to define internal energy, temperature, and pressure.

Thermohaline circulation (2) the buoyancy-driven part of the large-scale global ocean circulation. Different buoyancy (density) is created by the surface fluxes of heat (*thermo*) and salinity or salt (*haline*).

Tiling (7) a method for defining different fractional land-surface types where each type takes up only part of a grid box.

Time splitting (10) a method of model integration by which processes are split in time and each operates one after the other on updated model conditions from the previous process.

Time steps (4) discrete units of time that a model uses to integrate forward. Time steps are the time intervals for integration in a model.

Topography (7) the terrain or elevation of the earth's surface.

Transpiration (7) the process of evaporation of water from plant leaves. Formally, the process by which moisture is carried through plants from roots to small pores on the underside of leaves, where it changes to vapor and is released to the atmosphere.

Transport (5) the movement of a substance or characteristic such as temperature.

Tropical cyclones (1) the general term for a cyclone or storm over the oceans. The term includes tropical depressions, tropical storms, hurricanes, and typhoons. Tropical cyclones are classified according to their intensity: (1) tropical depression, with winds up to 17 m/s (38 mph); (2) tropical storms, with winds of 18–32 m/s (39–72 mph); and (3) severe tropical cyclones, hurricanes, or typhoons, with winds greater than 33 m/s (74 mph).

Tropopause (5) the top of the troposphere. The boundary between the troposphere and the stratosphere.

Troposphere (5) from the Greek words meaning "changing region." The region of the atmosphere up to 40,000 ft (12 km) or so. The region where most weather and clouds occur. Temperature decreases with height in the troposphere, making it often unstable and prone to vertical motion.

Uncertainty (1) the noncorrectable or unknown part of the inaccuracy of an instrument, system, or model. Uncertainty represents the limit of measurement (or forecast) precision.

Unstructured grids (6) a grid with irregular connectivity between elements and often elements of variable size.

Upwelling (11) ascending motion of subsurface water from deeper layers brought into the surface layer and removed from the area of upwelling by divergent horizontal flow.

Validation (5) the determination of how well a task is performed.

Variance (1) in statistics, variance measures how much a set of numbers is spread out. Zero variance means all values are identical. Variance is always positive: A small variance indicates that data cluster close to the mean. A high variance means that the data points are spread out from the mean (and each other).

Water table (7) the level below which the ground is saturated with water.

Weather (1) the state of the atmosphere, typically referenced to the surface of the earth, and characterized by different types of weather phenomena.

Weather forecasting (1) an estimate of the future state of the atmosphere, or future weather, usually defined in terms of temperature, winds, clouds, and precipitation.

Weathering (2) breaking down by exposure to weather (rain, freezing, water).

Wind stress (6) a force exerted on the ocean by the motion of the wind over it.

Index

© The Author(s) 2016
A. Gettelman and R.B. Rood, *Demystifying Climate Models*,
Earth Systems Data and Models 2, DOI 10.1007/978-3-662-48959-8